A Voyage up the River Amazon

Including a Residence at Pará

By William H. Edwards

The Narrative Press
True First-Person Historical Accounts

To Henry Longfellow Norris, Esq., of Para, this volume is most respectfully inscribed by the Author.

The text for this book was obtained from an original edition published in New York: 200 Broadway, by D. Appleton & Company, and Philadelphia: Geo. S. Appleton, 148 Chesnut St. MCDDDXLVII.

The Narrative Press
P.O. Box 2487, Santa Barbara, California 93120 U.S.A.
Telephone: (800) 315-9005 Web: www.narrativepress.com

©Copyright 2004 by The Narrative Press
All Rights Reserved. No part of this publication may
be used or reproduced in any manner whatsoever
without the prior written permission of the publisher.

ISBN **1-58976-244-4** (Paperback)

Produced in the United States of America

CONTENTS

Preface 1

Chapter I 3
Leave New York for Pará—Sunset—Curiosities of the sea—Luminous water—Approach the mouth of the Amazon—Salinas—Entrance of the river—Scenery—Arrival at Pará

Chapter II 7
Morning view of the harbor and city—Visit—Land at the Punto de Pedras—Novel scene—Reception at Mr. Norris's—Garden and plants—Electrical eel—Anaconda—Religious procession

Chapter III 12
Founding of Pará—Late disturbances—Site and vicinity—Form of the city—Rosinhas—Houses—Largo da Palacio, da Polvora da Quartel—Public buildings—Churches—Palaces—Theatre—Cathedral—Rua da Mangabeiras—Nazaré—Mr. Henderson's plantation—Rosinha of Mr. Smith, and fruit trees—Coffe—Pime-apples—Oranges—Limes—Mangoes—Inga—Alligator pears—Custard apple—Flowers

Chapter IV 21
License of residence—Officials—Provincial government—Church establishment—Troops—Enrollment of Indians—Drilling recruits—Absence of inns—Foreigners—Citizens—Manner of living—Public ball—Mechanics—Obstructions to labor—Apprentices and school—Carrying burdens—Water jars—Rearing of children—Food of lower classes—Mandioca and preparation of farinha—Tapioca—Fish—Beef—Vegetables—Fruits—Pacovas—Cocoa-nuts Assai palms

Chapter V 28
Leave Pará for the Rice Mills—Boatmen—Night scene upon the river—Arrival—Vicinity of the mills—A Brazilian forest—Sporting—Toucans—Chatterers—Motmots—Manikins—Illumming-birds—Snake stories—Absence of flies—

Ants—Saübas—Cupims—Little Ant-eater—Lakes—Nests of Troopials—Sloth—Armadillo—Beetles—Puma—Monkeys—Indian boy—Description of the mills—Blacks—Sleeping in hammocks—Vampire bats—Wasps' nests—Visit Corentiores—Sporting there—Reception—Bread fruit—Larangeira—Cotton tree—Maseranduba or Cow tree—Walk through the forest to the city—Spider—Flowers

Chapter VI 48
Start for Caripé—Island scene—Arrival—Vicinity—Tomb of Mr. Graham—Dinner—Shelling in the bay—Varieties of shells—Martins—Terns—Nuts and fruits—Mode of fishing—Four-eyed fish—Ant tracks—Moqueens—Forest—Creeping plants—Wild hogs, or Peccaries—Traps—Agoutis—Pacas—Squirrels—Birds—Chapel and singing of the blacks—Andiroba oil

Chapter VII 56
Leave for Taüaü—Indians—Arrival at midnight—Morning view—The estate—Tilaria or Pottery—Lime kiln—Slaves—Castanha tree—Cuya or Gourd tree—Ant hills—An ant battle—Forest—Macaws—Doves—Other birds—Sloth—Coati——Macura—Butterflies—Return to the city—Festival of Judas—Visit Sr. Angelico, upon the Guamá—Brazilian country house—Curious air-plant—Seringa or Rubber trees—Harpy Eagle—Monkeys

Chapter VIII 67
Leave Pará for Vigia—Boatmen—Inland passage—Egrets and herons—Stop at sugar plantation—Cupuassu—Mangroves—Insolence of pilot—Vigia—Arrival at Sr. Godinho's—Reception—The Campinha and its scenery—Sporting—Parrots—Employees—Sun-bird—Boat-bill—Tinami—Iguana lizard—Sugar cane—Mill—Slaves—Leave the Campinha—Kingfishers—Go below for Ibises—Sandflies—Return to Pará—A pet animal

Chapter IX 78
First discovery of the Amazon by Pinzon—Expedition of Gonzalo Pizarro—Descent of Orellana—Settlement of Pará—Second descent—Ascent of Teixera, and arrival at

Quíto—He descends with Acuña—Indian tribes—Rivers, etc.—Their report of the country—Number of tribes—Indian customs—Languages—Lingos Geral—Cannibals—System of the Jesuits—Their banishment—Present system, and condition of the Indians—The government—Compulsory labor

Chapter X 85
Preparation for ascending the Amazon—Our companions—The galliota—Indians—Provisions—Difficulties at starting—Detained at Sr. Lima's—Incident—An afternoon upon the beach—Another sitio—Marajo—The Tocantins—Islands—Ciganas and other birds—Wood scene—Habits of our Indians—Arrive at Braves—Pottery painting—Water-jars—Filing the teeth—Funeral of a child—A palm swamp—Seringa trees and gum collectors—Sloth—Howling monkeys—An adventure—Enter the Amazon—A macaw hunt

Chapter XI 96
Arrive at Gurupá—Situation of the town—Reception by the Commandant—An egg hunt—Storm—Cross the Xingu—Cárapanás—Cedar logs—Harpy Eagle—Birds—Mountains—Indian cooking—Forest trees—Snake birds—A Toucan's nest—Mutúcas—Indian improvidence—Grass fields—Enter an Igaripé—Hyacinthine Macaws—Passion flowers—Pass Pryinha—Monte Alégre—Arrive at Sitios—Thrush—Campo—Incident—Enter the Tapajos—White Herons—Flowering trees—Arrival at Santarem—Capt. Hislop—Morning calls—Beef—River Tapajos—Feather dresses—Embalmed heads—Description of Santarem—Departure—A slight difficulty

Chapter XII 108
The Amazon thus far—A cacao sitio—Politeness—Runaways—Growing of cacao—An alligator—High bank—Deserted sitio—Kingfishers—Romanças—Water birds—Arrive at Obidos—Rio des Trombetas—Incidents upon leaving—Manner of ascending the river—Shells—Stop at a sitio— High bluff—Water plants—Capitan des Trabalhadores—Arrive at Villa Nova—Fiesta of St. Juan—

Water scene—A Villa Nova house—Turtles—Stroll in the woods—Lakes

Chapter XIII 120
Leave Villa Nova—Our manner of living—Shells—Jacamars—Paroquets—Monkeys—Scorpion—Enter an igaripé—A deserted sitio—Wild duck—Scarlet Tanagers—A deserted sitio—Tobacco—Shells—A colony of monkeys—A turtle's revenge—Immense trees—Albino monkey—A self-caught fish—Porpoises—Curassows and nests—A turtle feast—Squirrel—Wild Indians—White herons—Shells—Umbrella chatterer—Cross to the northern shore—Periecu and Tambaki—Arrive at Serpa—Sr. Manuel Jochin—An Indian dance

Chapter XIV 132
Fourth of July at Serpa—Lake Saracá—An accession—Picnic—An opossum—Narrow passage—Swallow-tailed hawks—Sitio of the Delegarde—River Madeira—Village of our Taúcha— Appearance of his party on arriving at home—The old rascal—Bell-bird—Stop at a sitio, and reception—Orioles—A cattle satio—Swift current—Enter the Rio Negro—Arrive at Barra

Chapter XV 141
Rio Negra at Barra—The town—Old fort—Sr. Henriquez and family—Manner of living—Venezuelans—Piassába rope—Grass hammocks—Feather work—Descent of the Negro—Gallos de Serrs—Chili hats—Woods in the vicinity—Trogons—Chatterers—Curassows—Guans—Parrots and Tuucans—Humming Birds—Tiger Cats—Squirrels—A Tiger story—The Casuéris—A Yankee saw-mill—Mode of obtaining logs—A picnic—Cross the river to a campo—Cattle and horses—A select ball

Chapter XVI 152
A new river—Rio Branco—Turtle wood—Unexplored region—Traditions—Peixe boi or Cow Fish—Turtles—Influences at Barra—Indians—Foreigners—Indian articles—Poison used upon arrows—Traffic—Balsam Copaivi—Salsa—Quinia—Vanilla—Tonga beams—Indigo—Guaraná—

Pixiri or nutmeg—Seringa—Wild cotton—Rock salt—The Amazon above the Rio Negro—The Rio Negro

Chapter XVII 164
Prepare to leave Barra—Difficulty in obtaining men—The mail—Kindness of our friends—Re-enter the Amazon—Arrive at Serpa—A desertion—Working one's passage—Disorderly birds—Pass Tabocal—Snake-bird—Marakong Geese—Breeding place of Herons—Arrive at Villa Nova—The commandant—Visit to the Lake—Boat building—Military authorities—School—King of the Vultures—Parting with Sr. Bentos—Pass Obidos—Caracara Eagle—Our crew—Indian name of the Amazon

Chapter XVIII. 175
Arrive at Santarem—Negro stealing—Pass Monte Alegre—Strong winds—Usefulness of the Sun-bird—Family government—Reformation in the Paroquets—Low shore—A Congress—Otters—Enter the Xingu—Gurupá—Leave the Amazon—Assai palms—A friend lost and a friend gained—Braves—Our water jars—Crossing the bay of Limoeiro—Seringa trees—A lost day—Town of Santa Anna—Igaripé Merim—Enter the Moju—Manufacture of rubber shoes—Anatto—Arrival at Pará

Chapter XIX 186
Our Lady of Nazareth—Nazaré legend—Procession—Commencement of the fiesta—A walk to Nazaré—Gambling—Services in the chapel—An interesting incident

Chapter XX. 191
Leave Pará for Marajo—Voyage—Cape Magoary—Islands—A morning scene—Arrive at Jungual—A breakfast—Birds—Vicinity of Jungcal

Chapter XXI 196
Description of Marajo—Cattle—Tigers—Alligators—Snakes—Antas—Wild ducks—Scarlet Ibises—Roseate Spoonbills—Wood Ibises—Other birds—Island of Mixiana—Indian burial places—Caviana—Macapá—Bore or Pororoca—Leave Jungcal for the rookery—A sail among

the trees—Alligators—The rookery—Return—An alligator's nest—Adieu to Jungcal—Violence of the tide—Loading cattle—Voyage to Pará

Chapter XXII 206
Want of emigrants and laborers—Inducements to settlers, and disadvantages—Citizenship—Import and export duties and taxes—Want of circulating medium—Embarrassments of government—Capabilities of the Province—Effect of climate on the whites—the blacks—Inducements to the formation of a steamboat company—Seasons—Temperature—Health—Superior advantages to invalids—Farewell to Pará—Voyage home

PREFACE

In these stirring times, when all Anglo-Saxondom is on the qui-vive for novelty, and the discovery of a new watering-place is hailed with more enthusiasm than the discovery of a new planet;—when the "universal Yankee nation" has so nearly exhausted all the whereabouts which modern facilities for locomotion have brought so conveniently within its reach;—when the Old World has become also an old story, and Summer excursions to St. Petersburg and Tornea, and Winter sojourns in Australia and Typee, have afforded amusement, not only to travelers themselves, but to those who, at their own fire-sides, like equally well to take a trip to the ends of the Earth in their comfortable arm-chairs; it has been a matter of surprise to me, that those who live upon the excitement of seeing and telling some new thing, have so seldom betaken themselves to our Southern continent.

Promising indeed to lovers of the marvelous is that land, where the highest of Earth's mountains seek her brightest skies, as though their tall peaks sought a nearer acquaintance with the most glorious of stars; where the mightiest of rivers roll majestically through primeval forests of boundless extent, concealing, yet bringing forth the most beautiful and varied forms of animal and vegetable existence; where Peruvian gold has tempted, and Amazonian women have repulsed, the unprincipled adventurer; and where Jesuit missionaries, and luckless traders, have fallen victims to cannibal Indians, and epicurean anacondas.

With a curiosity excited by such wonders, and heightened by the graphic illustrations in school geographies, where men riding rebellious alligators form a foreground to tigers bounding over tall canes, and huge snakes embrace whole boats' crews in their ample folds; the writer of this unpretending volume, in company with his relative, Amory Edwards, Esq., late U.S. Consul at Buenos Ayres, visited Northern Brazil, and ascended the Amazon to a higher point than, to his knowledge, any American had ever before gone.

As an amusement, and by way of compensation to himself for the absence of some of the monsters which did not meet his curious eye, he collected as many specimens in different departments of Natural History as were in his power, at the same time chronicling the result of his observations, in the hope that they might not be unacceptable to the naturalist or to the general reader.

To the science or a naturalist he makes no pretensions, but as a lover, and devout worshiper of Nature, he has sought her in some of her most secret hiding-places, and from these comparatively unexplored retreats, has brought the little which she deigned to reveal to him.

The country of the Amazon is the garden of the world, possessing every requisite for a vast population and an extended commerce. It is, also, one of the healthiest of regions; and thousands who annually die of diseases incident to the climates of the North, might here find health and long life.

If this little book shall contribute to a more general knowledge of the advantages of such a country, the labor of its preparation will be amply repaid.

New York, May, 1847.

Chapter I

LEAVE NEW YORK FOR PARÁ—SUNSET—CURIOSITIES OF THE SEA—LUMINOUS WATER—APPROACH THE MOUTH OF THE AMAZON—SALINAS—ENTRANCE OF THE RIVER—SCENERY—ARRIVAL AT PARÁ

It was a cold morning, the 9th of February, 1846, that we left New York, in the bark Undine, Capt. Appleton, for Pará. Our fellow-passengers were Mr. Smith, the U.S. Consul of that port, his lady, and two young gentlemen, in quest, like ourselves, of adventures. Scarcely out of flight of Sandy Hook, a furious northwester burst upon us, and, for a week, we dashed on before it, at a rate to startle a landsman, had not the accompanying motion speedily induced that peculiar state, in which one would as lief not be, as be, and inclined to consider a bed beneath the waters as preferable to present torture. But the golden-haired spirit at the prow always smiled hopefully, and gallantly the noble bark sped onward to calmer waters and warmer skies. Here the sea was all loveliness, and, night by night, the scantily appareled sky of the north was disappearing before the as steadily advancing brilliance of the tropics. We watched the gradual descending of the north star; and when at last it sank below the horizon, it seemed as though an old and familiar friend had deserted us,—one whose place was not to be supplied even by the splendor of the southern cross.

By the twentieth day, we were near land, to the eastward of Salinas, having seen and enjoyed the usual sea-sights. Most memorable of these was a sunset, as we lay becalmed. The few snow-piled clouds that rested upon the water, gradually became suffused with flame, and the sea's surface was a sheen of green and gold, varying from one color to the other, as the rolling of the vessel changed our angle of view. A vapor fringe of rainbow hues circled the horizon, more lovely because rapidly changing, and beheld, as it were, through an atmosphere of floating golden particles. One by one the stars peeped out, and we fancied that we could detect a shade of sadness over their beautiful faces at having come too late.

We had seen sharks and brilliant-robed dolphins. A grampus had risen under the bow, and flying-fish had repeatedly flown on board. Many an hour we had whiled in fishing up gulf weed, and in observing the different species of animals with which it was filled.

As we neared the equator, the water became luminous; the waves were crested with fire; the vessel's path was one broad track of light, and as we took our shower bath under the pump, liquid flames dashed over us, and every drop was a splendor. To heighten our interest in the phenomenon, a score of porpoises were playing about in every direction, their tracks a living flame, contorted, zigzag, like fiery serpents. Now they would shoot out, rocket-like, leaving trains of thirty feet; now, darting back pursue each other round and round, till their path appeared a tangled skein of light.

The blue had changed to green; and long before land was visible, the green had lost itself in the muddy brown of the Amazon. Every where were discernible currents, known from afar, by their different hues, and by the furious boiling of their surfaces. Old Ocean was battling with the King of Rivers. Tossed about in the commotion were vast quantities of drift wood, fruits and plants. Huge fish-hawks were lazily flapping along. Gulls and terns were screaming.

In the night, a number of beautifully marked moths, attracted by our lights, visited us, and soon after daybreak, an inquisitive humming-bird came for a peep at the strangers, flitted about us a little time, then darted away to his home.

Salinas is an island at the mouth of the river, conspicuous from a distance, owing to its broad, white beach. It is principally inhabited by fishermen. We observed a few red-tiled houses, and an ancient white church. Here, vessels bound to Pará usually take a pilot; but owing to the vexatious delays often experienced, American captains prefer trusting to their own skill. Directly at the entrance of the river are two banks, Braganza and Tigoça, dreaded by sailors; beyond these, the navigation is easy. Pará is situated about eighty miles above; but such is the force of the descending tide and current, that from twenty-four to thirty hours are frequently required to overcome the short distance.

It was delightful to find ourselves once more in quiet water, and a luxury only appreciable by those who have been rolled and pitched about, until every bone seems rheumatic, and every muscle jelly-like, to sleep as stilly as on land. We had anchored inside the banks: before daybreak, we were again advancing; and, that morning, every passenger was early upon the look-out. The speedy termination of the voyage put us all in high spirits, and impatiently we snuffed the perfumed air that came wafted from the yet scarce visible shore. The island of Marajo gradually became distinguishable on the right, its tree tops but just fringing the water. To the left, long, low islands extended to within a few miles of the city. All day, our course was near these, and to one never before conusant of tropical luxuriance, and a truant from the wintry skies of the north, every thing was enchanting.

Impervious as a hedge, tall trees shot up their arrow-like stems; broad palm leaves undulated with every breath. A thousand shades of green were enamelled with flowers, in red, and white, and gold. The loud notes of the toucans, the shrill cries of parrots greeted our welcome; and about the vessel, twittered delightedly numbers of martins, the same old friends who used, at home, to disturb us in the early morning. Here and there, little patches of clearing, and haystack-shaped huts, indicated the home of some ease-loving Indian. Some of these huts consisted merely of a few poles, covered with palm thatch, but occasionally, a delicious little retreat would peep at us through the almost concealing shrubbery, surrounded by a grass-plot, and overshadowed by the huge leaves of the banana, or the feathery tufts of the cocoa tree. In front of one hut, upon a grassy knoll facing the river, stood a large cross, designed to warn away any evil spirit that should venture there. Happy ones! none but fairies, and good angels, should be welcome to such a paradise.

Often we saw men and women, walking upon the beach, or variously employed, and it was amusing to observe their pantomimic movements. Huge canoes, hollowed from single trees, and with mat sails, crept along shore: and the first strange voice that we had heard since leaving New York, hailed us from one of these, with the friendly "O Amigo."

Twenty miles below the city, a number of islands are sprinkled about the channel, one of which was pointed out as the last resort of the inhabitants of Pará, when the city was sacked by the rebel Indians, a few years since. Upon that lovely spot of green, five thousand persons died of exposure and starvation.

Pará is situated upon a little bay, forming a safe anchorage, and is visible, from below, a little more than ten miles. At about that distance, is the Quarantine, not now a terror to travelers. Here, a little boat, rigged with two antique triangular sails, and manned by negroes bare to the waist, pulled alongside, and left with us a custom-house guard, who was to prevent intercourse with the shore.

Night was coming on, but still there was light enough to display to our eager eyes, the position of the city, nestled in its bed of green, and smiled upon by an archipelago of islands. Rain commenced pouring, and we were fain to go below. The guard at the fort bid us pass on, and, by eight, we were anchored off the custom-house. It was too late for a visit, and we turned in, impatient for the morning. All night long, church bells were ringing, and clocks striking, and, at intervals, we could distinguish the notes of a bugle, or the loud cry of the patrol; all doubly cheerful, after the mournful wailing of the wind through the rigging, and the monotonous dashing of the sea, which had been our melancholy lullaby, for so many weeks.

Chapter II

MORNING VIEW OF THE HARBOR AND CITY—VISIT—LAND AT THE PUNTO DE PEDRAS—NOVEL SCENE—RECEPTION AT MR. NORRIS'S—GARDEN AND PLANTS—ELECTRICAL EEL—ANACONDA—RELIGIOUS PROCESSION

We had arrived in the midst of the wet season, and, all night, the rain poured incessantly. But, as the sun rose, the clouds broke away, and our first view was rendered still more agreeable by the roseate mist that draped the tree tops and lingered over the city. Anchored about us, were vessels of various nations and strange looking river craft, under whose thatched roofs, whole families seemed to be living, and, upon which, green parrots and macaws were clambering and screaming.

Canoes, bound to the market, were constantly passing, loaded with all kinds of produce. Fine looking buildings, of three and four stories height, faced the water, all yellow in color, and roofed with red tiles. Vast cathedrals and churches, covered with the mould of age, shot up their tall spires, their walls and roofs affording sustenance and support to venerable mosses and shrubs of goodly size. Garden walls were overhung with creeping vines, like ancient ruins. Vultures were leisurely wheeling over the city, or, in clusters, upon the housetops, spreading their wings to the sun. Mid the ringing of bells and the discharge of rockets, a long procession was issuing from the church of San Antonio; and a Babel of sounds, from dogs and parrots, and strange tongues, came over the water.

At about nine o'clock, the doctor of the port visited us; and soon after, an official of the custom-house examined our passports, and left with each of us a notification to present ourselves, within three days, to the chief of police, and to obtain from him a license of residence. We were then pronounced at liberty to go on shore.

It was low tide, and as no wharves run out for the convenience of vessels, we were obliged to land at the marketplace, the Punto de Pedras, a long, narrow pier. It would be impossible to conceive a more utterly novel tableau than here broke upon us. It was an introduction, at once, to half that was curious in the city. Files of canoes skirt the whole length of the pier, high and dry above the water. The more fortunate occupants, who have sold their wares, are variously engaged: some, sleeping; others, preparing their morning meal; others, combing and arranging their luxuriant tresses—for even an Indian woman has a little vanity; and others, the most of all, chattering with their neighbors, or screaming in shrill tones to friends on shore. Here are negroes of every shade of color, from the pure Congo, to the almost pure white; some buying, some selling. There stands one, with his basket of coarse cotton cloth and his yard-stick; and close by, an old wench is squatted by a pot of yellow soup, the extract of some palm nut. Here are strings of inviting fish, and piles of less captivating terrapins; coarse baskets, filled with Vigia crabs, the best in the world; and others of palm leaves, fashioned like a straw reticule, are swelled out with the delicious snails. Monkeys, fastened to clogs, entice you to purchase them by their antics; and while herons, and various other wild birds, by their beauty. Every where, and most numerous of all, are the fruit-dealers; and for a mere nothing, all the luxuries of this fruit-prolific clime are yours. Beautiful bouquets of flowers invite a purchaser; and now, for the first time, you observe the singularly neat appearance of the women, each dressed in white, and with a flower in her hair, and you remember that it is a holiday. Oddly-dressed soldiers mingle among the crowd; inquisitive officials peer about for untaxed produce; sailors, from vessels in the harbor, are constantly landing; gentlemen of the city are down for their morning stroll; beautiful Indian girls flit by, like visions; and scores of boys and girls, in all the freedom of nakedness, contend with an equal number of impudent goats, for the privilege of running over you.

Through this motley assemblage we picked our way, accompanied by Captain Appleton, to the house of Mr. Norris, the consignee of the Undine. Mr. Norris received us with all the warmth of an old friend, and immediately insisted upon our making his house our home. It was a home to us during our stay at Pará; and the generosity of Mr. N. has placed us under obligations easily understood by those, who, like ourselves, have found a home and a friend among strangers.

Our first excursion extended no further than the garden, at the rear of the house; but even that little distance opened to us a new world. It was laid out in home style, with neat walks and raised flower-beds. A number of curious birds were skulking among the shrubbery, or stalking along the path with the dignity and self-possession of birds at home. This domestication of wild birds, we afterwards found to be common throughout the province. They are restrained from truancy by the high fences that surround the gardens: and ibises and spoonbills, varieties of herons, rails, *et multi alii*, are as frequently seen as domestic fowls. But the legitimate occupants were of greater interest than these strangers: and here grew in perfection, the banana, the orange, the fig, the tamarind, the cotton tree, the sugar cane; and over the fence, on the soil of a neighbor, a lofty cocoa tree displayed its clusters of ripening nuts. Instead of the puny sensitive-plant, that, in the north, struggles almost hopelessly for frail existence, a giant shrub threw out its nervous arms, all flowering, and the attraction of passing butterflies.

Amid this profusion, there was nothing to remind us of the home that we had left; but, afar off, in one lone corner, stood a solitary stalk of Indian corn, lank and lean, an eight feet spindling, clasped nervously by one sorry ear. Poor thing, it spoke touchingly of exile.

Passing out of the garden, our next visit was complimentary to an eel: not one of the unhallowed denizens of muddy ponds, or stagnant waters; but an electrical eel, large and handsome, swimming about in his tub of clear rain water, with .the grace of a water king. This fellow was about four feet in length, and along his whole lower part extended a wide fin, by whose curvings he appeared to propel himself. We often, afterwards, amused our leisure in observing this eel, and in experimenting upon his electrical power. This did not seem to be concentrated in any particular part, or organ, for touch him where we would, the violence of the shock seemed the same, and equaled an ordinary shock from a machine. When very hungry, or particularly spiteful, he would transmit his power through the water to a considerable distance. His usual food was crabs, and when these were thrown in to him, he swam towards them, stunned them by a touch of his head, and either caught them immediately, or allowed them to fall to the bottom of the tub, to be devoured at leisure.

These eels are common in the small streams about Pará, and, indeed, throughout the whole northern part of the continent, and they often attain great size. One that we afterwards saw at Senhor Pombo's, was about six feet long, and five or six inches in diameter. We heard frequent accounts of their power over large animals in the water. The negroes catch them by first teasing them, until they have exhausted the electrical power. We ate of them, at different times, but they were too fishy in taste to be agreeable, without strong correctives.

Near by, was disclosed to us a young anaconda, nicely coiled up in the bottom of a barrel, and looking as innocent as a dove. This fellow was pointed out as something rather diminutive, but to our unfamiliar eyes, a snake of ten feet length seemed very like a monster. His customary food was rats. These snakes are kept about many houses in Pará for protection against rats, and two who had escaped from Mr. Norris's barrels, now prowled at large, and effectually cleared the premises of these vermin. They are perfectly harmless, and never molest domestic fowls or animals upon the premises, excepting, now and then, a young chicken.

This day was a festival. The saint was popular, business was suspended, public offices were closed, and the whole city was preparing to do him honor. Such days, in Pará, always end in processions, and when, late in the afternoon, the crackling of rockets, and the sounds of martial music, proclaimed the procession already formed, we walked to the Rua da Cadeira, the Broadway of Pará, and took our stand among crowds of citizens, all, apparently, as much interested as ourselves in the coming events. The balconies above were filled with gaily dressed ladies, and bright eyes were impatient to pay their homage to the benignant saint, or to exact a homage, more sincere, perhaps, from their own admirers below.

Immediately succeeding a fine military band, walked a number of penitents, wearing crowns of thorns, and almost enshrouded in long, black veils. It was evident enough that peccadilloes were not all confined to the whites, for, below the veils, bared feet displayed as many hues as we had seen in the marketplace. These penitents surrounded a tall banner, borne by one of their number, who staggered beneath its weight; a fair penance for many a hearty sin.

Friars, with corded waists and shaven crowns, and priests, in long black robes, came next. Little angels followed, bright, happy things, and beautiful, as though they had come down to cheer the present sufferings of the weary one, who bore his cross behind. Each wore upon her head a crown of flowers, and exquisite devices decked her white gauze dress. Wings of a butterfly, or some shorn Cupid, told bow she came; she bore a wine cup in her hand, and as she stepped, tiny bells sent out low music. She was unaccustomed to our rough walks here, and, at her side, a seraph boy guided her faltering steps.

Then came the Christ, bending beneath the heavy cross. The crowd was stilled, the Host passed by, and respect, or adoration, were testified by raised hat, or bended knee.

A number of other figures succeeded, and the line was closed by the troops. A few whites followed, curious as ourselves; but the whole negro and Indian population were drawn along, as a matter of course. Nearly all the negro women were profusely ornamented with gold, partly the fruit of their own savings, and often, the riches of their lady mistresses, who lend them willingly upon such occasions. Some wore chains of gold beads, passing several times about the neck, and sustaining a heavy golden cross. All wore ear-rings, and the elder women, both black and Indian, overtopped their heads by huge tortoise-shell combs. The Indian girls, who were in large numbers, were almost always beautiful, with regular features, fine forms, black, lustrous eyes, and luxuriant locks, that fell over their shoulders. Many women carried upon their heads trays, covered with a neat towel, and well provided with temptations to errant coin.

At intervals along the street, were little buildings, in which temporary altars were fitted up in all the glare and gaudiness of wax candles and tinsel. Every one raised his hat upon passing these, and the more devout knelt before them, depositing some coin at their departure.

In the evening, the churches were brilliantly lighted, and in the alcoves, before the images of the saint, knelt crowds of ladies, the elite of Pará. At each altar priests officiated, their attention much distracted between the fair penitents at their side, and the dulcet tones in the money plate before them.

Another procession, by torch-light, closed the exercises, and at last, wearied with sight-seeing, we wended our way homeward, to the embrace of luxurious hammocks, that gently received us, without the usual misadventure of the uninitiated and uncautioned.

Chapter III

FOUNDING OF PARÁ—LATE DISTURBANCES—SITE AND VICINITY—FORM OF THE CITY—ROSINHAS—HOUSES—LARGO DA PALACIO, DA POLVORA DA QUARTEL—PUBLIC BUILDINGS—CHURCHES—PALACES—THEATRE—CATHEDRAL—RUA DA MANGABEIRAS—NAZARÉ—MR. HENDERSON'S PLANTATION—ROSINHA OF MR. SMITH, AND FRUIT TREES—COFFE—PIME-APPLES—ORANGES—LIMES—MANGOES—INGA—ALLIGATOR PEARS—CUSTARD APPLE—FLOWERS

The popular name of this city, Pará, is derived from the river, its proper designation being Belem, or Bethlehem. Caldeira, in 1615, entered what he supposed to be the main Amazon, and learning from the natives that this was, in their language, the King of Waters, called it, appropriately, Pará; or rather, to hallow it by a Christian baptism, the Gram Pará. Continuing up the river, this adventurer at last fixed upon a site, near the junction of several streams, now known as the Guamá, the Acará, and the Mojú, for a city, that should thereafter be a glory to our Lady of Belem. Our Lady is still the patron saint, but the name of her city is almost entirely forgotten in that of Pará.

We will not recount the long series of events that have transpired since Caldeira here first planted the cross. They would be of little interest to the general reader, and we prefer to look at the city as it now is, merely making such allusions to the past, as shall serve to render description more intelligible.

The only event that requires particular mention, is the Revolution of 1835, and the following year. The President of the province was assassinated, as were very many private individuals of respectability, and the city was in possession of the insurgent troops, assisted by designing whites and Indians. All the citizens who could, fled for their lives; many to Portugal, and many to the United States and England. The whole province, with the exception of the town of Cametá, upon the Tocantins, fell into the hands of the rebels, and every where, the towns were sacked, cities despoiled, cattle destroyed, and slaves carried away. The rebels were constantly quarreling among themselves, and several Presidents succeeded each other. At last, after this state of anarchy had continued nearly eighteen months, President Andrea arrived from Rio Janeiro with a sufficient force, and succeeded, without much difficulty, in recovering possession of the city. One by one, the inland towns returned to their allegiance. The disastrous effect of these disturbances is still felt, and a feeling of present insecurity is very general, but still, Pará has fully recovered her former position, and may retain it, if the provincial government guides itself with sufficient discretion.

The whole Amazonian region is low, and the site of the city boasts no advantage in this respect, being, at most, but a few feet above the level of the river at flood tide. Every where, nature displays the most exuberant fertility, and this, which, in most countries between the tropics, is a prolific source of pestilence and death, is here so modified by other elements as to be a blessing. During the rainy season, when, for several months, rain falls daily, and for several weeks, almost incessantly, the surface of the ground is never long covered with water; for, so sandy is the soil, that, no sooner have the clouds broken away, than the waters have disappeared, and excepting the bright jewels that sparkle profusely upon every leaf, little else remains to tell of the furious outpourings of the previous hour. During what is termed the dry season, from June to December, more or lees rain falls weekly, and vegetation is never disrobed of her perennial green. The steady trade winds from the East come fraught with invigorating sea air, tempering the fierce sun-heat, making the nights of a delightful coolness, and preventing that languor of feeling so inseparable from the equatorial climes of the East.

Old traditions, handed down as applicable to modern times, by all-knowing Encyclopedists, represent the climate of Pará as having been unhealthy, but in some respects improved of late years. These reports probably arose from the injudicious method of living introduced by the earlier colonists, and persevered in, until experience taught them to accommodate their habits to the clime. But, of late years, they have been studiously detailed and exaggerated by monopolizing mercantile houses; and when we desired to venture to the country of the Amazon, it was next to impossible to obtain any sort of information relative to Pará, except a general report of heat and unhealthiness. I shall speak more of this hereafter, with reference to the singularly superior advantages which Pará presents to invalids.

The whole city is laid out in squares, and, from the peculiar manner in which it is built, covers a much larger area, than, from its population of fifteen thousand, one would suppose. Near the river, and in the part more especially devoted to business, the houses adjoin, upon streets of convenient width; but elsewhere each square is usually the residence of but one proprietor, who here enjoys all the advantages of both city and country. These residences are termed *rosinhas*. Fruit trees, of every variety common to the clime, mingle with beautiful flowers, and it requires but little taste in the master or ladies of the mansion to embower themselves in a paradise. Most of these houses are but of one story, built upon two or three sides of a square, covering a great area, and containing numerous lofty and well ventilated rooms. Very often, the entire flooring is of neat, square tiles. A broad verandah offers both shelter and shade, and here, in delicious coolness, the meals of the day are enjoyed.

The city proper consists of houses of every height, from one to four stories, strongly resembling each other in external appearance. All are yellow-washed or white-washed, and ornamented by mouldings about doors and windows. The building materials are small stones cemented in mortar, and such is the durability of construction, that unfinished walls, in different parts of the city, exposed, for years, to the action of the elements, show no sign of crumbling or decay. Of course, coolness is the great object aimed at, and therefore, in the centre of the house is usually an open square from top to bottom, serving to keep up a constant current of air. Doors are all wide, and windows rarely glazed. Generally, near the river, the lower part of the house is occupied as a store or ware-room, the upper stories being the residence of the family.

In front of upper windows opening upon the street are iron balconies, favorite stands of the inmates, who here spend hours, in the cooler parts of the day, in observing the passers below, and sometimes, it is to be feared, coquetting with correspondents over the way. It strikes one strangely that necessity has not introduced the fashion of shaded balconies as a protection from the sun; but there are none such, and in positions sheltered from the sea breeze, the mid-day heat is excessive.

The lower houses, in the more retired streets, are mostly dwellings, and the windows of these are always covered by a close lattice, or jalousie, through whose bars dark eyes may flash upon passers-by unblushingly.

The streets are without sidewalks, and are badly paved with irregular stones, which render walking excessively fatiguing, and rapid riding perilous.

In different parts of the city, are public squares, called Largos. The more prominent are the Largo da Palacio (of the palace); da Polvora (of powder); and da Quartei (of the barracks). The first of these is very spacious, and might be made an ornament to the palace and the city. As it is, it is neither more nor less than a dirty common, uneven in surface, spotted, in the wet season, with puddles of water, and unshaded by a single tree. Miserable, half-starved sheep, parti-colored as goats, and libels on the ovine race, glean a poor subsistence from the coarse rank grass. The walk across this Largo to the palace was of rough stone, and when we first crossed it, both daylight and dexterity were requisite; but I am happy to say, that, before we bade adieu to Pará, preparations were making for an avenue more consistent with the dignity of the Government.

Upon the Largo da Polvora formerly stood the powder-house, now removed to a distance from the city. Here trees were once planted by President Andrea, but with merely exceptions enough to show what a public blessing their preservation would have proved, they have now disappeared. Near this Largo, are the principal wells, whence is supplied the water for the city, and about which, may be seen, at any time, scores of negro women, engaged in washing and bleaching clothes.

The Largo da Quartel is of small extent, fronting the barracks, a long, low building, where Indian recruits are drilled into civilization and shape. In the centre of this Largo, is a well, about the curb of which, numbers of considerate wenches rest their weary water-jars, and with a painful self-denial, gossip and gesticulate, all day long, upon the affairs of the town.

The public buildings of Purá are conspicuous objects, both in number and size far beyond the present wants of the city; but wisely built for posterity, and the future inevitable magnitude of the depot of the Amazon. Even so long ago as 1685, when the population numbered but five hundred, there existed "a Mother Church, a Jesuit College, a Franciscan, a Carmelite, and a Mercenario Convent, two Churches, a Chapel, and a Misericordia or Hospital." The cherished hopes of the Jesuits have not yet been fulfilled, but" already is heard the sound of the multitude that is coming to take possession of the valley."

The Jesuit college has now become an ecclesiastical seminary; and the convents, long since deserted of friars, save two or three old Franciscans, have been turned to profaner uses. That of the Carmelites, is now the palace of the assembly; the vast pile of the Mercenaries has become the custom-house; and still another is the arsenal. All these edifices are in good preservation, and the bright green moss, which every where has climbed the roofs, and traced the facings, in no wise detracts from their picturesque appearance.

The palace, built about the middle of the last century, when Portugal looked to the Amazon as the scene of her future glory, is commensurate, in size and massiveness, with the anticipated necessities of the empire. It is of the same style of architecture as the Portuguese houses generally, and can scarcely be called either grand or beautiful.

In the rear of the palace, stands the unfinished theatre, now overgrown with shrubs and close embracing vines; a far greater ornament to the city, than it could have been in its finished state.

The cathedral stands near the palace, upon the southern side of the Largo; the vastest edifice of the kind in Brazil. Twin steeples tower aloft, from whose many bells issue most of those chimes, that may be heard at almost any hour.

Near the arsenal, and sufficiently removed to be no nuisance to the city, is the public slaughter-house, where are received all the cattle destined for the Pará market. Strangers usually walk in that direction, to observe the immense congregation of vultures that are here to be seen, laboring lustily for the public health.

There are a number of pleasant walks, within and around the city. The most agreeable, by far, of the former, is the Rua da Mangabeiras, a long avenue, crossed, at right angles, by a similar rua, and both thickly skirted by mangabeira trees. This tree attains a vast size, and throws out a more widely spreading top than most Brazilian forest trees. Its bark is a singular combination of colors, between green and gray; and is of a lustrous smoothness. The ripened fruit hangs over the branches; large red pods, the size of a cocoa-nut, and containing a yellowish, silky cotton. In the months of March and April, these trees are divested of their leaves; and every where mingle in profusion, the ripened fruit, and the large, white, crown-like flowers. Later in the season, the flowers have given place, in turn, to a most luxuriant foliage; and when the sun strikes mercilessly upon every spot else, here, all is coolness and repose. Paroquets, ravenously fond of the cotton seeds, are every where chattering among the branches; and the brilliant cicadas chirp grateful thanks to him who planted for them this delightful home. From adjacent thickets, come the warblings of many birds; and the stranger, haply unacquainted with the Brazilian melodists, startles, as he hears the liquid trill of the blue bird; the joyful song of the robin, and the oriole's mellow whistle. 'Tis a delusion; but the familiar tones sound none the less delightfully, from the throats of these southern cousins, than when uttered amid the groves and by the streams, of our own home.

The Rua da Mangabeiras is deservedly a favorite walk in summer, and in the early morning, or after sunset, it is constantly thronged with groups of joyous citizens.

Another delightful walk, as well as the usual route for equestrians, is towards Nazaré, distant about two miles from the palace, and one mile from the city. Here is a little chapel dedicated to the service of our Lady of Nazareth, and looking like some fairy's palace, on its spot of green, embowered in the native forest. Our Lady of Nazareth is the peculiar patroness of the sick, the afflicted, and the desolate; and here, the soul-saddened penitent may find quiet, far away from the crowded shrines of the city. At the entrance of the square, a number of seats invite the weary. A tall, white pillar, standing near, records, probably, some event connected with the place, but the inscription is nearly illegible.

With our friend Captain Appleton, who is a most zealous conchologist, and well acquainted with all the shell-haunts in the vicinity, we used often to take this route, and upon the trees, in various localities, found as many specimens as we cared for. These were principally of three varieties: the Bulimus regius, Bulimus glabra, and the Auricula clausa. Continuing on through the forest, at about a mile beyond Nazaré, is the plantation of Mr. Henderson, a Scotch gentleman, who, having a taste for agricultural pursuits, is endeavoring to show the planters of the country the difference between a scientific cultivation, and their own slovenly and inefficient mode of farming. Amongst other novelties, Mr. H. has introduced a plough, the only one in the province of Pará. He has devoted particular attention to the cultivation of grasses for hay, and his meadows looked as freshly, and produced as fine grass as those of New England. What with the delightful reception of Mr. Henderson, and the lesser attractions of scenery and flowers, butterflies and shells, we took many a stroll this way.

But there was no pleasanter place, wherein to while an hour, than a rosinha, and as our friend, Mr. Smith, was proprietor of one of the most extensive, within a ten minutes' walk of our residence, we used often to visit him, and amuse ourselves among his trees. This rosinha was of about an acre's extent. Down the middle ran a broad walk, covered by an arbor, which was profusely overrun by the Grenadilla passion-flower. This produces a yellow fruit, about the size and shape of an egg, within which is a pleasant acid pulp.

On either side the arbor were coffee trees. These are planted at a distance of about ten feet apart, and being prevented from growing more than five feet high, by constant trimming of their tops, they throw out very many lateral branches. The flowers are white, and, at the flowering season, ornament the plant beautifully. The leaves are about six inches in length, broad, and of a rich and glossy green. The berries grow upon the under side of the limbs, and at first, are green, but when matured of a deep red. Within each are two kernels, and the whole is surrounded by a sweet, thin pulp. When the ripe berries are exposed to the sun, this pulp dries, and is then removed by hand, or by a mill. The trees produce in two or three years after being planted. Formerly the quantity of coffee raised in the vicinity of Pará was sufficient for a large exportation, and it was celebrated for its superior flavor. Now it is imported, so many planters having turned their attention to other produce, or to the collecting of rubber.

There were also large patches of ananas, or pine-apples, which plant is two well known to require description. This fruit is often raised in these rosinhas, of great size. One which we saw upon the table of the British Consul, soon after our landing, weighed nineteen pounds, and was considered nothing extraordinary, although, at that time, out of the season.

A number of large orange trees were always interesting to us, inasmuch as, at every season, they clustered with ripe fruit, not the shriveled or sour specimens seen in New York, but of great size and luscious sweetness. Oranges, in this climate, are to be considered rather as a necessity, than a luxury. Their cooling nature renders them unspeakably grateful, and they are, without doubt, an antidote to many diseases incident to a torrid clime. Every one uses them unstintingly, and when an old gentleman, upon the Upper Amazon, told us that he always settled his breakfast with a dozen oranges, he described, with little hyperbole, the custom of the country.

There were also many lime trees; and these resemble, in general appearance, the orange, excepting that they are of smaller growth. The acid of limes is more pleasant than vinegar, and they are always used as a substitute for this upon the table. They are much used in composing a drink, and make the best of preserves.

The most beautiful trees were the mango and the ochee, whose densely leaved tops much resemble each other. Their leaves are very long and narrow, and of a dark, glossy green; but when young they are of several shades, dull white, pink, and red, and the commingling of hues is very beautiful. The mango is esteemed one of the finest fruits. It is the size of a large lemon, and of a green color. Beneath the skin is a yellow pulp, which surrounds a large stone. During our stay mangoes were temporarily unpopular among the lower classes, from a belief that to them was owing the appearance of a disease called the leprosy.

The ochee is smaller than the mango, and of a yellow color. It contains a sweet, pleasant pulp.

Another interesting tree was the ingá, although for a very different reason than its beauty. It bears a profusion of small, white flowers, very fragrant; and the attraction of humming-birds, who might, at any time, be seen rifling their sweets, in a great variety of species. The fruit of the ingá is a pod, of a foot or more in length, and an inch in diameter. It contains a sweet, white pulp, imbedded in which are long seeds. The paroquets are very fond of this pulp, and they come to the trees in great flocks, clustering upon the pods, and tearing them open with their strong beaks.

There were trees bearing another esteemed fruit, the alligator pear, or mangába. Of these there are two varieties, one, the more common, green on color, and shaped like a crook-necked squash, but of greatly reduced size. The other, considered the better species, is called the mangába da Cayenne and is of the ordinary pear shape, and of a purplish red color. In the centre is a large stone, and the substance about this is soft and marrow-like. It is eaten with wine and sugar, and to our taste was the finest fruit in the province. It is said to be the only fruit that cats will eat, and they are extremely fond of it.

The birabá, or custard-apple, is no bad representative of the delicacy of which its name is suggestive. It is about the size of a cocoa-nut, covered by a thin, rough skin, and contains a white pulp, which is eaten with a spoon.

Here was growing a cactus, in size a tree; and numerous flowering shrubs, some known to us as green-house plants, and other entirely new, were scattered over the premises. Cape jessamines grew to large shrubs and filled the air with fragrance. Oleanders shot up to a height of twenty feet, loaded with flowers; and altheas, in like manner, presented clusters of immense size and singular beauty. Here, also, was a tree covered with large, white flowers, shaped like so many butterflies; and there were a host of other, of which we could admire the beauty, although not knowing the names.

Chapter IV

License of residence—Officials—Provincial government—Church establishment—Troops—Enrollment of Indians—Drilling recruits—Absence of inns—Foreigners—Citizens—Manner of living—Public ball—Mechanics—Obstructions to labor—Apprentices and school—Carrying burdens—Water jars—Rearing of children—Food of lower classes—Mandioca and preparation of farinha—Tapioca—Fish—Beef—Vegetables—Fruits—Pacovas—Cocoa-nuts Assai palms

Within the three days limited in our notification, we had called upon the chief of police for a license of residence, which was furnished us gratuitously. This officer was one of the many examples that we met with, of the disregard paid to color, in public or private life, throughout the country. He is considered the second officer of the Provincial Government, and, like the President, receives his appointment directly from Rio Janeiro.

In passing our chattels through the custom-house, also, we had not experience the least difficulty or annoyance, the officers discharging their duties in the most gentlemanly manner. And, at all times, in our intercourse with officers of the Government, we found them extremely polite and obliging, and generally, they were men of intelligence and education.

The President, with three Vice-Presidents, constitute the Executive of the Province. Assemblies of deputies, chosen by the people, meet at stated seasons at Pará, to regulate provincial matters. They have a greater license, in some respects, than the corresponding branched of our State Governments, such as the imposing of tariffs, and the like, but their acts are referred to Rio Janeiro for confirmation.

The Judges of the various districts, who are also chiefs of police, are appointed at Rio, but the Justices of the Peace are chosen by the people.

The church establishment of Pará is not large, when the wants of the whole province are considered; but, as by far the large portion of the padres never go beyond the city, their number seems disproportionate. One meets them at every step, and probably five hundred is not an exaggeration. Of these, many are novitiates in different stages of preparation, and the grades are readily distinguished by their differences of dress. Since convents have become unpopular, the old race of friars have almost disappeared; still, a few are seen, and a small number of others are among the Indians of the interior. The clergy are, of course, very efficient patrons of the three-and-thirty holidays, besides divers festivals extraordinary, that diversify the Brazilian year.

Near the Ecclesiastical Seminary is the school for young ladies, under the supervision of the sisters of some of the religious societies. Here a great number of young ladies from various parts of the province receive education in the simpler branches, and in what would be called "the finishing" of a New York boarding-school.

The Catholic is the established religion of the state, but all religions are tolerated. There is no other sect in Pará, and probably within the province, out of the city, preaching of any other denomination was never heard.

The regular troops of the empire are collected in this province in great strength, on account of the revolutionary spirit of the people. Every morning they are paraded upon the Largo da Palacio until eight o'clock, and then marched down the Rua da Cadeira to the music of a fine band. They are out upon every public occasion, taking part in every procession. They are, moreover, the police of the city, and in discharge of their duties, are seen scattered, throughout the day, along the pier and streets, and guarding the door, of all public offices. Night police, as well as day police, they take their stations, in the early evening, about the city, and, at every hour, their loud cries disturb the sleepers.

Upon Sundays, these troops are freed from duty, and the National Guard take their places, on parade or at the sentry. This Guard, one would suppose, formed a far more efficient force than the regular army; the one, composed, as it is, of native Brazilians, the other, a heterogeneous compounding of white and black, yellow, red, and brown. The Indian seems to predominate, however, and it might be questionable how far his courage would carry him, once led into action.

During the last few years, the enrollment of Indians has been carried to an unprecedented extent, through apprehension of renewed disturbances. Since 1836, ten thousand young men are said to have been carried to the south, to the incalculable injury of the agricultural interest. As might be supposed, all this enlistment has not been voluntary. The police are constantly upon the alert for recruits, and, the instant that a poor fellow sets foot within the city, he is spirited away, unless some protecting white is there to intercede in his behalf. We frequently fell in with cottages in the vicinity of the city, whose only occupants were women and children, the men having, in this way, disappeared. Most of the market boats, also, are managed by women, the men often stopping at some convenient place above, and there awaiting the boat's return.

It is an amusing sight to watch these Indian recruits, during their earlier drillings, upon the Largo; encumbered with oppressive clothes, high leathern stocks beneath their chins, and a wilderness of annoying straps about their bodies. Their countenances are models of resignation, or of apathetic indifference, when the drill-officer has his eye upon them; but when that eye is averted, the nervous twitching, and the half-suppressed curses, with which they wipe the beaded sweat from their brows, would be ludicrous enough, could one overcome a feeling of pity at the predicament of the poor devils.

Free negroes are very apt to be caught in the same trap; and then, negroes and Indians, together, spend their leisure hours, off drill, in the lock-up; until, between the principles of honor therein imbibed, and the ardor of military glory excited, they can be considered trustworthy, and suffered to go at large. Most free negroes avoid this career of greatness, by nominally still belonging to their old master, or some other willing protector.

There are no inns, at Pará, for public accommodation. The people from the country do not require them; each having friends in the city, or conveniences for living on board his vessel. Strangers visiting the port are usually provided with introductory letters to some of the citizens, and are received with the most generous hospitality. There are various cafés, where a good cup of coffee or chocolate may always be obtained; but these are not very much patronized. Both natives and foreigners, engaged in business, provide at their own tables, for their clerks, or others connected with them in business; a system productive of mutual advantages.

A great proportion of the foreigners in the city, are from the United States and Great Britain; and these form among themselves a delightful little society.

The people of the town are native born Brazilians and Portuguese; often well educated, generally intelligent, and always polite. Of the lower classes, very many are Portuguese or Moorish Jews, who obtain a livelihood by trafficking with the smaller river craft, by adulterating produce, and by various other expedients in which the people of that nation are expert.

Most gentlemen residing in the city, have also estates in the country, to which they retire during summer. Their mode of living is very simple, and in congeniality with the clime. Two meals a day, are considered quite sufficient; and late suppers are entirely avoided.

Most of the business of the day is transacted in the early morning; and when the noon's heat is beating, "all," as they say, "but Englishmen and dogs," are taking a siesta in their hammocks. The cool evening, lovely and brilliant, calls out every one; and a round of pleasure encroaches far into the night. Parties and balls are constantly being given; and all over the city is heard the light music of the guitar, and the sounds of the joyous dance. Upon the last Saturday evening of each month, is a public subscription ball, and Pará's beauties are there, in all the fascination of flashing eyes, and raven hair, and airy movements. Sometimes a theatrical company ventures into this remote region, and, for a while, the new prima donna is all the rage.

The mechanics of the city are mostly Portuguese, and have all the proverbial industry of their nation. A shoemaker, who lived opposite us, used to be rather annoying in this respect; pegging away at all hours of the night, and not sparing time to breathe, even on Sundays.

Owing to the imperfection, or entire absence of machinery, the labor of an artisan is far more toilsome than with us, and he compensates the difference, by something more than proportionate slowness. The cabinet maker has to saw his materials, from the log, in his own shop, and two or more boys, lazily pulling away at a pit-saw, are always a part of his fixtures. So with other trades. Such a state of things would be excessively annoying, any where else, but these people are accustomed to it, probably dream of nothing better, and are well content to jog on in the safe and sure path, by which their ancestors, God rest them, moved forward to glory.

There is this deficiency, throughout the province, with respect to every sort of labor-saving machinery; and although, now and then, some individual of extraordinary enterprise has introduced improvements from other countries, and although the government allows new patents of machinery to be entered without a duty, yet the mass of proprietors know nothing of them. The introduction of machinery would compensate, in a great degree, the depressing scarcity of laborers, for want of whom, this garden of the world lies desolate.

Very many of the apprentices in the shops are Indian boys, and to facilitate the acquisition of trades by these, the government supports a school, where, in addition to the common branches of education, fifty Indian boys are instructed in various trades. This institution owes its existence to President Andrea, who seems to have had concentrated in him, more benevolence and public spirit, than a score of those who preceded or succeeded him in office. It is to him, that the city is indebted for the Rua da Mangabeiras, and this alone should immortalize a man in Pará.

The absence of horses and carts, together with the universal custom of carrying burdens upon the head, seem, at first, an oddity to a stranger. In this manner, the heaviest as well as the lightest, the most fragile as well as any other, travels with equal safety to its destination. For the convenience of vessels there are two companies of blacks, each numbering thirty men, who are regular carriers; and their noisy cries are heard every morning, as in the full tide of some wild song, they trot off beneath incredible burdens.

Every where, are seen about the streets, young women, blanks or Indians, bearing upon their heads large trays of *doces*, or sweetmeats and cakes, for sale. These things are made by their mistresses, and are thus marketed. Nor do the first ladies of the city consider it beneath their dignity thus to traffic, and we heard of some notable examples, where the money received for the doces had accumulated to independent fortunes. From similar large trays, other women are huckstering every variety of vegetables or fruits; and not unfrequently meets the ear the cry of as-sy-ee, the last syllable prolonged to a shrill scream. What assai may be, we shall soon explain.

In a morning walk, in any direction, one encounters scores of blacks, men and women, bearing huge water jars to and from the different wells, which are the supply of the city. These jars are porous, and being placed in a current of air, the water attains a delightful coolness. This custom was borrowed by the early settlers from the Indians, and is universal. In various parts of the house are smaller jars, called bilhas (beelyas), by the side of which stands a large tumbler, for the general convenience.

The habit of carrying burdens upon the head, contributes to that remarkable straightness and perfection of form, observed in all these blacks and Indians. Malformation, or distortion of any kind, is rarely encountered. This is doubtless owing, in a great degree, to the manner of rearing children. Every where, are to be seen swarms of little boys and girls, unrestrained by any clothing whatever, and playing in the dirt with goat and dogs. This exposure to the sun produces its natural effect, and these little people, blacks and whites, are burned into pretty nearly the same tint; but they grow up with vigor of constitution and beauty of form. The latter, however, is sometimes ludicrously modified by a great abdominal protrudence, the effect of constant stuffing with farinha. It is very unusual to hear a child cry. The higher classes, in the city, are more careful of their children; but, in the country, the fashion of slight investment prevails, and, at the Barra of the Rio Negro, the little son and heir of the chief official dignitary was in full costume, with a pair of shoes and a cane.

The food of all the lower classes, throughout the province, consists principally of fish and farinha. The former is the dried and salted Periecu, of the Amazon; the latter, a preparation from the Mandioca root. This plant, botanically, is the Jatropha Manihot, known in the West Indies as Cassava. The stalk is tall and slender, and is divided into short joints, each one of which, when placed in the ground, takes root, and becomes a separate plant. The leaves are palmated, with six and seven lobes. The tubers are shaped much like sweet potatoes, and are a foot or more in length. They are divested of their thick rind, and grated upon stones; after which, the mass is placed in a slender bag of rattan, six feet in length. To this, a large stone is appended, and the consequent extension producing a contraction of the sides, the juice is expressed. The juice is said to be poisonous, but is highly volatile. The last operation is the drying, which is effected in large iron pans, the preparation being constantly stirred. When finished, it is called farinha, or flour, and is of a white or brown color, according to the care taken. In appearance it resembles dried crumbs of bread. It is packed in loose baskets, lined with palm leaves, and in the bulk of eighty pounds, or an *alquier*. Farinha is the substitute for bread and for vegetables. The Indians and blacks eat vast quantities of it, and its swelling in the stomach produces that distention noticed in the children.

Tapioca is made from the same plant, and is the starchy matter deposited by the standing juice.

The rivers are filled with varieties of fine fish, but, in the city, many other articles of diet are considered preferable. From Vigia, and below, towards the coast, crabs and oysters are brought, at certain seasons, in great abundance. The former, particularly, are noticeable for their large size and superior flavor; but the oysters, though of prodigious size, can, in no way, be compared with their relatives of the north. They are found, in large clusters, about the roots of the mangroves.

The great dependence or the Pará market ii beef. Upon Marajo, and neighboring islands, vast herds of cattle roam the campo, and large canoes are constantly engaged in transporting them to the city. But often, they are poor when taken, and the passage from the islands averaging from four days to a week, during which time, they have little to drink, and nothing at all to eat, those who survive are but skin and bone. Killed in this state, it may readily be imagined that Pará beef is deficient in some points, considered as excellencies in the Fulton market. It is cut up in shapeless pieces, without any pretence at skill. The usual method of preparing it for the table is to boil it, such a dish as legitimate roast beef or steak being unheard of.

Very few potatoes, of any sort, are seen; the principal vegetables for the table being rice, fried plantains, and an excellent variety of squash, called jurumu.

It is in fruits that Pará excels; and here is a long catalogue, many of which are common to adjacent countries, within the tropics, and many others peculiar to this province. Of many of these, we have already spoken; but there are two or three others, which deserve mention: and first of these are the plantain, and pacova, or banana. These fruits resemble each other, excepting in size; the former being of about eight inches length, the latter, in its varieties, from three to five or six. The producing tree is one of the most beautiful of the palms, the coronal leaves being six feet in length, by two broad, and gracefully drooping around the trunk. The fruit hangs in clusters about a stalk, depending from the top of the plant. While still green, the stalk is cut off, and the fruit is suffered to ripen in the shade. The plantains are generally prepared for eating, by being cut in longitudinal slices, and fried in fat; but when roasted in the ashes, are extremely pleasant, and reminded us strongly of roasted apples. The pacovas are eaten raw, and are agreeable and nutritious. They are raised without difficulty, from cuttings, and are the ever-present attendant of the gentleman's garden or the Indian's hut. Their yield, when compared with other plants, is prodigious, being, according to Humboldt, to wheat, as one hundred and thirty-three to one, and to potatoes, as forty-four to one.

Cocoa palms are abundant upon the plantations, and are conspicuous from their long, feather-like leaves, and the large clusters of nuts which surround their tops. The nuts are generally eaten when young, before the pulp has attained hardness.

From various palm fruits are prepared substances in great request among different classes of people; but, most delightful of all, is that from the Euterpe edulis, known as assai, or more familiarly, as, was-sy-eé. This palm grows to a height of from thirty to forty feet, with a stem scarcely larger than one's arm. From the top, a number of long leaves, their webs cut, as it were, into narrow ribbons, are waving in the wind. Below the leaves, one, two, and rarely, three stems put forth, at first enclosed in a spatha, or sheath, resembling woven bark. This falling off, there is disclosed a tree-like stalk, with divergent limbs, in every direction, covered with green berries, the size of marbles; these soon turn purple, and are fully ripe. Flocks of toucans, parrots, and other fruit-loving birds, are first to discover them; but there are too many for even the birds. The fruit is covered by a thick skin, beneath which, imbedded in a very slight pulp, is the stone. Warm water is poured on, to loosen the skin, and the berries are briskly rolled together in a large vessel. The stones are thrown out, the liquid is strained off the skins, and there is left a thick, cream-like substance, of a purple color. Sugar is added, and farinha to slightly thicken it. To a stranger, the taste is, usually, disagreeable, but soon, it becomes more prized than all fruits beside, and is as much a necessity as one's dinner.

Chapter V

Leave Pará for the Rice Mills—Boatmen—Night Scene upon the River—Arrival—Vicinity of the Mills—A Brazilian Forest—Sporting—Toucans—Chatterers—Motmots—Manikins—Illumming-birds—Snake Stories—Absence of Flies—Ants—Saübas—Cupims—Little Ant-eater—Lakes—Nests of Troopials—Sloth—Armadillo—Beetles—Puma—Monkeys—Indian boy—Description of the Mills—Blacks—Sleeping in Hammocks—Vampire Bats—Wasps' Nests—Visit Corentiores—Sporting there—Reception—Bread fruit—Larangeira—Cotton Tree—Maseranduba or Cow Tree—Walk through the Forest to the City—Spider—Flowers

Our first excursion, to any distance, was to the Rice Mills, at Magoary, only twelve miles from Pará by land, and two tides, or about ten hours by water. The overland route being, in many respects, inconvenient, we determined to venture in one of the canoes, always in readiness for such excursions, near the Punto da Pedras; and for this purpose, engaged a fair looking craft, with a covered and roomy cabin, and manned by two whites and a negro. Leaving the city in the middle of the afternoon, we took advantage of the ebbing tide, and, by dart, had entered the stream, which was to carry us to our destination. But our two white sailors were lazy scoundrels, and we did not feel sufficiently acquainted with the language, or accustomed to the ways of the country, to give them the scolding they deserved. This they knew enough to comprehend, and the consequence was, that we lost the flood tide which should have carried us up, and were obliged to anchor and spend the night on board. One of these men was an old salt, battered and worn, the other was a young fellow of twenty, with a good-looking face and nut-brown skin, wearing upon his head a slouched felt hat, and, altogether, the very image of peasant figures seen in Spanish paintings. Not at all disturbed by our dissatisfied looks, and ominous grumblings, they coolly stretched themselves out upon the seats, and started up a wild song, the burden of which was of love, and the dark-eyed girls they had left behind them in the city. It was a lovely night, and the music, and other gentle influences, soon restored our good humor, and we felt, at last, inclined to forgive the laziness that had led us here. No clouds obscured the sky, and the millions of starry lights, that, in this clime, render the moon's absence of little consequence, were shining upon us in their calm, still beauty. The stream, where we were anchored, was narrow; tall trees drooped over the water, or mangroves shot out their long finger-like branches into the mud below. Huge bats were skimming past, night-birds were calling in strange voices from the tree-tops, fire-flies darted their mimic lightnings, fishes leaped above the surface, flashing in the starlight, the deep, sonorous baying of frogs came up from distant marshes, and loud plashings in shore, suggested all sorts of nocturnal monsters. 'Twas our first night upon the water, and we enjoyed the scene, in silence, long after our boatmen had ceased their song, until nature's wants were too much for our withstanding, and we sank upon the hard floor to dream of scenes far different.

It was eight o'clock in the morning, when turning an angle of the stream, we came full in view of the mill, the proximity of which we had been made sensibly aware of, for the last half hour, by the noisy clamor of the machinery. It was a lofty stone structure, standing forth in this retirement, like some antique erection. Mr. Leavens was expecting us, and we were delighted once more to shake the hand of a warm-hearted countryman. Breakfast was upon the table, and here, for the first time, we ventured to test our capacities for fish and farinha. The fish was a hard case, coarser than shark meat, and requiring an intimacy with vinegar and oil to remove its unpleasant rankness. Farinha was not so disagreeable, and we soon came to love it as do the natives. Indeed, long before our Amazonian experience had ended, we could relish the fish, also, as well as any Indian.

The scenery about the mill is very fine. In front, the stream, a broad lake at high water, and a tiny brook at other times, skirting a low meadow, at the distance of a hundred rods, is lost in the embowering shrubbery. All beyond is a dense forest. Upon the meadow, a number of large, fat cattle are browsing on the coarse grass, and flocks of Jacanas, a family of water-birds remarkable for their long toes, which enable them to step upon the leaves of lilies and other aquatic plants, are flying with loud cries from one knoll to another. Back of the mill, the road leads towards the city, and to the right and left are well-beaten paths, leading to small, clear lakes, from which the mill derives its water. The whole vicinity was formerly a cultivated estate, but the grounds are now densely overgrown. At the distance of a mile, the road crosses what is called the first bridge, which spans a little stream that runs sporting through the woodland. The color of the water of this, and other small streams, is of a reddish cast, owing, doubt-less, to the decomposing vegetation. It is, however, very clear, and fishes, and eels, may at any time be seen playing among the logs and sticks which strew the bottom. Beyond this bridge is the primeval forest. Trees of incredible girt tower aloft, and from their tops one in vain endeavors to bring down the desired bird with a fowling-piece. The trunks are of every variety of form, round, angular, and sometimes, resembling an open net-work, through which the light passes in any direction. Amid these giants, very few low trees or little underbrush interferes with one's movements, and very rarely is the path intercepted by a fallen log. But about the trees cling huge snake-like vines, winding round and round the trunks, and through the branches sending their long arms, binding tree to tree. Sometimes they throw down long feelers, which swing in mid air, until they reach the ground, when, taking root, they, in their turn, throw out arms that cling to the first support. In this way, the whole forest is linked together, and a cut tree rarely falls without involving the destruction of many others. This creeping vine is called sepaw, and, having the strength and flexibility of rope, is of inestimable value in the construction of houses, and for various other purposes.

Around the tree trunks clasp those curious anomalies, parasitic plants, sometimes throwing down long, slender roots to the ground, but generally deriving sustenance only from the tree itself, and from the air; called hence, appropriately enough, air-plants. These are in vast numbers, and of every form, now resembling lilies, now grasses, or other familiar plants. Often, a dozen varieties cluster upon a single tree. Towards the close of the rainy season, they are in blossom, and their exquisite appearance, as they encircle the mossy and leafed trunk, with flowers of every hue, can scarcely be imagined. At this period, too, vast numbers of trees add their tribute of beauty, and the flower-domed forest, from its many colored altars, ever sends heavenward worshipful incense. Nor is this wild luxuriance unseen or unenlivened. Monkeys are frolicking through festooned bowers, or chasing in revelry over the wood arches. Squirrels scamper in ecstasy from limb to limb, unable to contain themselves for joyousness. Coatis are gamboling among the fallen leaves, or vicing with monkeys in nimble climbing. Pacas and agoutis chase wildly about, ready to scud away at the least noise. The sloth, enlivened by the general inspiration, climbs more rapidly over the branches, and seeks a spot, where, in quiet and repose, he may rest him. The exquisite, tiny deer, scarcely larger than a lamb, snuffs exultingly the air, and bounds fearlessly, knowing that he has no enemy here.

Birds of gaudiest plumage, flit through the trees. The trogon, lonely sitting in her leaf-encircled home, calls plaintively to her long absent mate. The motmot utters his name in rapid tones. Tucáno, tucáno, comes loudly from some fruit-covered tree, where the great toucans are rioting. "Noiseless chatterers" flash through the branches. The loud rattling of the woodpecker comes from some topmost limb; and tiny creepers, in livery the gayest of the gay, are running up the tree trunks, stopping, now and then, their busy search, to gaze inquisitively at the strangers. Pairs of chiming-thrushes are ringing their alternate notes, like the voice of a single bird. Parrots are chattering; paroquets screaming. Manakins are piping in every low tree, restless, never still. Wood-pigeons, the "birds of the painted breasts," fly startled; and pheasants, of a dozen varieties, go whirring off. But, most beautiful of all, humming birds, living gems, and surpassing aught that's brilliant save the diamond, are constantly darting by; now, stopping an instant, to kiss the gentle flower, and now, furiously battling some rival humble-bee. Beijar flor, kiss-flower, 'tis the Brazilian name for the humming bird, beautifully appropriate. Large butterflies float past, the bigness of a hand, and of the richest metallic blue; and from the flowers above, comes the distant hum of myriads of gayly coated insects. From his hole in the sandy road, the harmless lizard, in his gorgeous covering of green and gold, starts nimbly forth, stopping, every instant, with raised head and quick eye, for the appearance of danger; and armies of ants, in their busy toil, are incessantly marching by.

How changed from all this, is a night scene. The flowers, that bloomed by day, have closed their petals, and nestled in their leafy beds, are dreaming of their loves. A sister host now take their place, making the breezes to intoxicate with perfume, and exacting homage from bright, starry eyes. A murmur, as of gentle voices, floats upon the air. The moon darts down her glittering rays, till the flower-enameled plain glistens like a shield: but in vain she strives to penetrate the denseness, except some fallen tree betrays a passage. Below, the tall tree trunk rises dimly through the darkness. Huge moths, those fairest of the insect world, have taken the places of the butterflies, and myriads of fire-flies never weary in their torch-light dance. Far down the road, comes on a blaze, steady, streaming like a meteor. It whizzes past, and, for an instant, the space is illumined, and dewy jewels from the leaves throw back the radiance, 'tis the lantern-fly, seeking what he himself knows best, by the fiery guide upon his head. The air of the night bird's wing fans your cheek, or you are startled by his mournful note, wac-o-row, wac-o-row, sounding dolefully, by no means so pleasantly as our whippoorwill. The armadillo creeps carelessly from his hole, and, at slow pace, makes for his feeding ground; the opossum climbs stealthily up the tree, and the little ant-eater is out pitilessly marauding.

All this supposes pleasant weather; but a storm in these forests has an interest, though of a very different kind. Heavy clouds come drifting from the east, preceded by a low, ominous murmur, as the big drops beat upon the roof of leaves. Rapidly this deepens into a terrific roar; the forest rocks beneath the fury of the blast, and the crashing fall of trees resounds fearfully. Tornadoes are unfrequent; but one, while we were at the mills, swept through the forest, now, hurling aside the massive trees like weightless things, and now, tripping carelessly, only taking tribute of the topmost boughs—sportive in its fierceness. We were struck by the absence of thunder and lightning in the furious pourings of the rainy season. The clouds came to their daily task gloomily, as though pining for a holiday, and, in the weariness of forced toil, forgot their wantonness.

Our first gunning expeditions were between the mill and the bridge, and the nature of the woods rendered it a toilsome matter, until experience had made us acquainted with the most convenient paths, and the notes and habits of the birds. Every one venturing into the forest is armed with a long, curved knife, called a *tresádo*, for the purpose of cutting his way through the entangling vines, that especially obstruct the woods of second growth. In such a section, also, the foliage is so dense, that it is extremely difficult to discover the birds who are uttering their notes all about—and when they are shot, it is often a puzzle to the keen eyes of an Indian to find them, amid the vines. But one soon learns that most of the families have peculiar haunts, where, early in the morning, or late in afternoon, they congregate in flocks. The trees in these places, are usually thickly covered with berries of some sort, and until these are entirely exhausted, the concealed sportsman may shoot at the perpetually returning flocks, until he is loaded with his game. Berries succeed berries, so constantly, throughout the year, that, in some spot, the birds' food is never wanting.

Most noticeable of all these birds, both for size and peculiarity of form, are the Toucans. There are many varieties, appearing at different seasons; but the Red-billed, R. erythrorynchos, and the Ariel, R. ariel (Vig.) are the largest and most abundant, seen at every season, but towards autumn, particularly, in vast numbers throughout the forest. Their large beaks give them a very awkward appearance, more especially when flying: yet, in the trees, they use them with as much apparent ease, as though they were, to our eyes, of a more convenient form. Alighted on a tree, one usually acts the part of sentinel, uttering constantly the loud cry, Tucáno, whence they derive their name. The others disperse over the branches, climbing about by aid of their beaks, and seize the fruit. We had been told that these birds were in the habit of tossing up their food to a considerable distance, and catching it, as it fell; but, as far as we could observe, they merely threw back the head, allowing the fruit to fall down the throat. We saw, at different times, tamed toucans, and they never were seen to toss their food, although almost invariably throwing back the head. This habit is rendered necessary, by the length of the bill, and the stiffness of the tongue, which prevents their eating as do other birds. All the time, while feeding, a hoarse chattering is kept up; and, at intervals, they unite with the noisy sentry, and scream a concert that may be heard a mite. Having appeased their appetites, they fly towards the deeper forest, and quietly doze away the noon. Often in the very early morning, a few of them may be seen sitting silently upon the branches of some dead tree, apparently, awaiting the coming sunlight before starting for their feeding trees.

The nests of the toucans are represented in works of Natural History, as being constructed in the hollows of trees. It may be so in many cases, and with some species. The only nest that we ever saw, which was of the Toco toucan, was in the fork of a large tree, over the water, upon the Amazon.

Toucans, when tamed, are exceedingly familiar, playful birds, capable of learning as many feats as any of the parrots, with the exception of talking. When turning about, on their perch, they effect their object by one sudden jump. They eat any thing, but are particularly fond of meat. When roosting, they have a habit of elevating their tails over their backs. The beaks of the Red-billed toucans are richly marked with red, yellow, and black; but preserved specimens soon lose this beauty. The other varieties found near Pará are the Pteroglossus maculirostris (Licht); the P. bitorquatis (Vig.); and the C. viridis. The family of birds most sought after by collectors, and the most gaudy of the Brazilian forests, is that of the Chatterers. There are several species, four of which are not uncommon in the vicinity of Pará, each about the size of the blue-bird of the United States. One of these, Ampelis Cayana, the Purple-throated chatterer, is of an ultramarine blue color, having a bright metallic luster, and with a throat of purple velvet. Another, A. cotinga, the Purple-breasted, is of a deep blue, similarly metallic, and ornamented both upon throat and breast with purple. A third, the White-winged, A. lamellipennis, is of a lustrous black, with wings and tail a snowy white. The fourth, A. carnifex, is with us called the Cardinal; in the language of Brazil, Passaro do sol, bird of the sun; and well be deserves the name. The crest upon his head is of scarlet, resembling the finest silk; his back and wings are of a golden bronze, and his tail and breast of the most delicate vermilion. All these birds may be seen at Mr. Bell's, our prince of taxidermists, and even when dimmed of their glories and encaged in glass, are pre-eminently beautiful. But when, in large flocks, they cluster in the tree-tops, dazzlingly lustrous in the sunlight, even the kiss-flower might be envious. These birds have no song. That charm, impartial nature has conferred upon others outwardly less attractive; and these must be content with a simple note. The Cardinal is lees common than the others, and is more generally seen in pairs, breeding in the months of August and September, near the mills. The other species seem transient visitors, generally abundant in May and June, and, at that season, associating in large flocks. There is another variety, the Carunculated chatterer, sometimes called the Bell-bird, occasionally seen near Pará. Mr. Leavens seems to be the only person who has met with them, having obtained a pair in the deep forest. This bird is the size of a small dove, and of a pure white color, when mature. On the bill is a fleshy caruncle, about an inch in length, somewhat like a turkey's comb. Of its habits or its note, we could learn nothing. The more common Chatterers are inactive birds, and great gluttons, often eating until quite stupefied. In this, they resemble their relative, the Cedar-bird of the north.

The Motmot, Momotus Brasiliensis, is another of these curious residents. This bird is about the size of a robin, having a back of a dark, rich green, and a long wedge-shaped tail, two feathers of which extend some inches beyond the others. The shafts of these are stripped of their webs near the extremities, giving the bird a very singular appearance. One would suppose that these birds trimmed their feathers thus themselves, for many are found with quills perfect, and others, partly denuded. The Motmots are generally in pairs in the deep woods, and are easily recognized by their note, motmot, slowly repeated.

The Manikins, in their different varieties, form a beautiful family; the most numerous of any, and corresponding much in their habits to our warblers. They are tiny things, generally having black bodies, and heads of yellow, red, white, and other colors. Like perpetual motion personified, they move about the branches and low shrubs, always piping their sharp notes; and unless upon a feeding-tree, almost defying shot.

The common varieties are the White-capped, Pipra leucocilla; Red-headed, P. erythrocephala; Blue-backed, P. pareola; and Puff-throated, P. manacus. Of these, the first is most abundant. A nest of the Red-headed was composed of tendrils of vines; and was scarcely larger than a dollar, and very shallow. It was affixed to one of the outermost forks of a low limb, beyond reach of any enemy but one. The eggs were cream-colored, and speckled with brown. A nest of the Blue-backed was composed of leaves, fibers, and moss, and much resembled in shape a watch-case. A nest of the Puff-throated was also pensile, but not so ingeniously composed as either of the others. The eggs of the two latter species were cream-colored and much spotted, particularly at the larger end.

Many other remarkable species of birds I shall have occasion to speak of hereafter; at present, I will mention but the humming-birds. Wherever a creeping vine opens its fragrant clusters, or wherever a tree flower blooms, may these little things be seen. In the garden, or in the woods, over the water, every where, they are darting about, of all sizes, from one that might easily be mistaken for a different variety of bird, to the tiny Hermit, T. rufigaster, whose body is not half the size of the bees buzzing about the same sweets. The blossoms of the inga tree, as before remarked, brings them in great numbers about the rosinhas of the city, and the collector may shoot, as fast as he can load, the day long. Sometimes, they are seen chasing each other in sport, with a rapidity of flight arid intricacy of path, the eye is puzzled to follow. Again, circling round and round, they rise high in mid air; then dart off like light to some distant attraction. Perched upon a little limb, they smooth their plumes, and seem to delight in their dazzling hues; then, starting off, leisurely they skim along, stopping capriciously to kiss the coquetting flowerets. Often, two meet in mid air and furiously fight, their crests, and the feathers upon their throats, all erected, and blazing, and altogether pictures of the most violent rage. Several times, we saw them battling with large black bees who frequent the same flowers, and may be supposed often to interfere provokingly. Like lightning our little heroes would come down, but the coat of shining mail would ward their furious strokes. Again and again would they renew the attack, until their anger had expended itself by its own fury, or until the apathetic bee, once roused, had put forth powers, that drove the invader from the field.

A boy in the city, several times, brought us humming-birds alive, in a glass cage. He had brought them down, while, standing motionless in the air, they rifled the flowers, by balls of clay, blown from a hollowed tube.

The varieties found about Pará are, principally, the White-collared, T. mellivorus; Hermil, T. rufigaster; Topaz-throated, T. pella; Tufted-necked, T. ornatus; Magnificent, T. magnificus; Scaly-back, T. eurynomus; Even-tailed amethyst, T. orthusa; Emerald, T. bicolor; Eared, T. auritus; Rough-legged racket-tail, T. Underwoodi; Sapphire-throated, T. sapphirinus; Violet fork-tail, T. furcatus; Sable wing, T. latipennis; Blue green, T. cyaneus. We received from Mr. Leavens a nest of the Hermit. It was formed upon the under side of a broad grass leaf, which drooped in a manner to protect it entirely from sun and rain. The material of which it was composed was a fine moss. Day after day, Mr. L. had watched its formation, hut before the little architect had completed it, the ants appeared, and she sought a safer spot for her home.

At first, we were somewhat nervous about venturing far into the woods, and anxiously careful to protect our feet from vicious reptiles by redoubtable boots. A little experience served to disabuse us of this error, and we were soon content to go in slippers. Old bugbear stories of snakes began to lose their force, when day after day passed without meeting even a harmless grass-snake. Not that there really are no such animals, for sometimes, huge specimens have been seen about the mills, and one, not many months before, had been surprised, who in his fright disgorged a fine musk-duck. But such cases are of extreme rarity, and only occur near the water. In the forest, snakes are not seen, and no one thinks of fearing them.

The absence of flies seems still more strange to a person from the North, who has always been accustomed to associate flies with warm weather, and who, mayhap, has been tormented by black swarms, in our woods. Their place, in Brazil, is well supplied by ants, who are seen every where, in the houses and in the fields. But as the main efforts of these insects are directed to the removal of whatever is noxious, most species are not merely tolerated, but looked upon as sincere and worthy friends. They are of all sizes and colors, from the little red fire-ant, who generally minds his own business, but who, occasionally, gets upon one's flesh, making all tingle, to the huge black species, an inch or more in length, who labors zealously in the woods for the removal of decaying vegetation. In this work, this ant is assisted by a smaller variety, also black; and armies, two and three feet wide, and of interminable length, are frequently encountered in the woods. It well becomes one to stand aside from their line of march, for they turn neither to the right nor to the left, and, in a moment, one may be covered to his dismay, if not sorrow.

But there is one variety of ant which must be excluded from all commendation. This a small species, called Saüba, and they are a terrible annoyance to the proprietors of rosinhas, inasmuch as they strip the fruit trees of their leaves. An army of these will march to the tree, part ascending, and the others remaining below. Those above commence their devastation, clipping off the leaves by large pieces, and those below shoulder them as they fall, and march away to their rendezvous. It is surprising what a load one of these little things will carry, as disproportionate to its size, as if a man should stalk off beneath an oak. Before morning, not a leaf is left upon the tree, and the unfortunate proprietor has the consolation of knowing, that unless he can discover the retreat of the saübas, and unhole them, one by one, every tree upon his premises will be stripped.

There is a small white ant called Cupim, that builds its nest in the trees, at the junction of a limb, or often, about the trunk. These are sometimes of great size, and, at a distance, resemble black knurls. Upon this variety the little Ant-eater lives. Climbing up some convenient tree, he twists his long, prehensile tail about the trunk, or some favoring limb, and resting upon this, commences operations. Making an incision in the exterior of the nest, by means of the sharp, hook-tike claws, with which his arms are furnished, he intrudes his slender snout, and long, glutinous tongue. So well protected by wool is he, that the ants have no power over him, but abide their fate. I kept one of these animals for some days, but he refused all nourishment. During the day, he sat with his tail twisted around a limb appropriated to his use, his head buried in his fore paws. But when the dusk of evening came on, he was wide awake, and passed half the night in walking pretty rapidly about the room, seeking some egress, and in climbing about the furniture. The negroes have a belief that if the Ant-eater is shut up in a tight box, and secured by every possible means, he will be spirited away before morning. The most intelligent black about the mills came to me, desiring I would try the experiment. "He is a devil," said Larry, and I consented, shutting his impship in a wooden chest. Next morning, Larry's eyes opened, as he saw the test had failed, and he signified his intention to believe no more lies, for the future.

The lakes, in the vicinity, were interesting places of resort to us, and several times we pushed the little canoe, or *montaria*, up the raceways, and paddled about amid the bushes, or along the shores, in search of birds or nests. The latter were very common, and it was interesting to observe the care with which the building spot was chosen, to keep it from the reach of lizards, or other reptiles, but above all, from the ever-present ants. And yet the ants were always there; they had passed from shore, upon leaves and floating shrubs, and every tree was infested by them. Most of the nests were arched over above, to keep out the sun's heat; and particularly those of the Flycatcher family, who, in the north, build open nests.

The most singular nests, and most worthy description, were those of the Troopials, Cassicus icteronotus (Swain), a large, black bird, much marked with yellow, and frequently seen in cages. Their native name is Japim. They build, in colonies, pensile nests of grass, nearly two feet in length, having an opening for entrance near the top. Upon one tree, standing in the middle of the lake, not more than ten feet high, and the thickness of a man's arm, were forty-five nests of these birds, build one upon another, often one depending from another, and completely concealing all the tree-top, except a few outermost leaves. At a distance, the whole resembled a huge basket. Part of these nests belonged to the Red-rumped Troopial—C. haemorrhous—and a singular variety of oriole, the Ruff-necked, of Latham, called Araona, or Ricebird, after the fashion of our Cow-bird, deposits its eggs in the Troopials' nests, leaving the young to the care of their fostermothers. Upon this tree, was a small hornets' nest, and the Indian whom we employed, asserted that these were the protectors of the birds from intruders. It may be so; we saw the same fellowship at other places. Usually, Troopials build nearer houses, and are always welcome, being friendly, sociable birds, ever ready to repay man's protection by a song. Often, in such situations, large trees are seen with hundreds of these nests dependent from the limbs, and swaying in the wind. A colony which had settled upon a tall palm, near the mill, was, one night, entirely robbed of eggs by a lizard. Snakes are sometimes the depredators, and between all their enemies, the poor birds, of every species, are robbed repeatedly. Probably owing to this cause, it is very unusual to find more than two eggs in one nest.

The Red-rumped troopials shot in this place, were of different sizes, some being several inches longer than others, although all were in mature plumage. Their nests were perhaps larger than those of the Japim's, but differed in no other respect. The eggs were white, spotted with brown, and particularly on the larger end. The Japim's eggs were cream-colored, and similarly spotted; and the eggs of the Ruff-necked orioles were large in proportion to the size of the bird—bluish in color, and much spotted, and lined with dark-brown.

We employed an Indian who lived near by, by name Alexandro, and a notable hunter, to obtain us specimens and to serve as guide, upon occasions. He never could be induced to shoot small birds, but always made his appearance with something that he considered legitimate game—often a live animal. One of these captives was a Sloth, and this fellow we kept for several days, trying to see what could be made of him. He was a pretty intractable subject, and poorly repaid our trouble. In face he resembled somewhat a monkey, and the corners of his mouth curving upward, gave him a very odd appearance, making him look, as one would suppose a monkey toper might look, if monkeys ever dissipated. His long arms were each terminated by three large claws, and his tough skin was well protected by a shaggy coat of coarse, grisly hair. Placed upon the ground, he would first reconnoiter, turning his head slowly about, then leisurely stretch forth one arm, endeavoring to hook his claw in something that might aid him in pulling himself onward: this found, the other claws would slowly follow, in turn. He uttered no noise of any kind. But put him where there was opportunity to climb, and his appearance was different enough: that dulled eye would glisten, and an idea seem to have struck him; rapidly his arms would begin to move, and sailor-like, hand over hand, he would speedily have climbed beyond recovery, had not a restraining rope encircled him. These animals are very common through this forest; but upon the Amazon, far more numerous. There are certainly two very distinct varieties, and the Indians say three. Usually, they are seen upon the lower side of a horizontal limb, hanging by their curved claws. They sometimes eat fruit, but principally live upon leaves; and when these are stripped from one tree, betake themselves to another, which they in turn denude.

At another time, Alexander brought in a young Armadillo, or Tatú, which he had dug from its burrow in the ground. There are several varieties about Pará. They are easily tamed, eating all sorts of vegetables, and insects, particularly beetles, which they unhole from their hiding-places in the earth. I went, one day, with Alexander, to the margin of one of the lakes in the woods, to obtain specimens of a coveted beetle, (Phanæus lancifer). We found a number of their holes, reaching down to the level of the water, rather more than two feet. Fragments of wing-cases of the beetles were strewed about, and many holes, of a larger size, explained that the Tatú had been before us.

In one of Alexander's excursions, he had the good fortune to discover a full-grown Puma, in the act of devouring a deer, which it had just killed. Nothing daunted, although armed with but a single-barreled gun, and that loaded with BB shot, he gave the animal a discharge, which made him leave the deer, and spring to a tree. Six several times our hunter fired, until, at last, the Puma was dead at his feet. Formerly, these animals were not uncommon, but now, are very rarely met, except upon Marajo.

Not unfrequently the fruit of our hunting excursions was a Monkey, and we considered this most acceptable, as it furnished our table with a meal, delicious, though not laid down in the cookery books. These animals are eaten throughout the province, and are in esteem beyond any wild game. Whatever repugnance we felt at first, was speedily dissipated, and often, in regard to this as well as other dishes, we had reason to congratulate ourselves, that our determination of partaking of whatever was set before us, discovered to our acquaintance many agreeable dishes, and never brought us into trouble.

Somewhere in these precincts, A—— picked up a little naked Indian, with eyes like a hawk, and most amusingly expressive features. Squatted upon a bench, with his knees drawn up to his chin, he would watch every motion with the curiosity of a wild man of the woods. A—— denominated him his tiger, but the black servitors shook their heads, and muttered "un poco diablo," a little devil. It was the tiger's business to follow in the woods, and pick up game, and in the intricacy of a thicket, rarely could even a hummer escape him. Here he was at home; but in the house, the indistinctness of his conceptions of meum and tuum, and his ignorance of the usages of even a tolerably decent society, made him very annoying. One day, being rated for not having dried A——'s shirt, he was discovered, soon after, with the shirt upon his back, and standing over the fire.

The building, a part of which as now used as a rice mill, was formerly appropriated to different purposes, and was the manor house of a vast estate, now mostly unproductive. It was in the days of Pará's glory, under the old regime, and here, upon the finishing of the structure, were gathered all the beauty and aristocracy of the city—coming down in barges, with music and flying streamers, to a three days' revel. Every Sunday, the old proprietor rode through the forest to the city, with coach and four. Those days have passed, and the boundless wealth and the proud aristocracy that surrounded the viceroy's court, have passed with them. An American company formed at Northampton, Mass., purchased the estate, and, for many years, under the superintendence of Mr. Upton, the agent and main proprietor, have carried on a large and profitable business. There are two mills, one propelled by stream, the other by water. The rice is brought in canoes from the city, and being hulled, is returned, to be reshipped, in great part, to Portugal. In this level country, it is extremely difficult to find a sufficient of water for a mill seat, but still more so, to find a fall so conveniently situated as to be accessible by tide water. Both these requisites are here; the fall of water being twelve feet, and the flood tide filling a deep basin directly by the side of the mill. About twenty blacks are employed upon the place, and the more intelligent are found every way competent to attend the different departments. Larry, particularly, was a general favorite with visitors, and had showed his appreciation of their favor, by picking up a few words of English. His province was filling and marking the sacks, and being paid a price for all above a certain number, he earned, regularly, between two and three dollars a week. We thought, of course, that Larry was in a fair way to he a freeman, and, in our innocence, suggested that he was laying up money to buy his papers. But he dispersed all such notions by the sententious reply, "I do not buy my freedom, because I am not a fool." He had a good master, he had a wife, and he did not have care or trouble. Thus he was contented. The aspirations of another of these blacks, were more exalted; for one day, as he sat ruminating upon air castles, his soul fired, perhaps, with the glorious "excelsior," he burst out with, "I wish I was a rich man, I would eat nothing but fresh fish." The wood used in the steam mill was brought up by canoes, and exchanged for broken rice. It was handsome split wood, tough as hickory, and of varieties generally capable of a fine polish. Most of those who brought it were women, and they threw it out and piled it, as though they were not unaccustomed to the labor. There was one little boy, of not more than nine years, who used to paddle, alone, a small montaria, unload his wood, buy his rice, and return with the tide. This was nothing unusual, but it serves to show the confidence reposed in children, who, at an early age, are often seen in situations thought to require double the years elsewhere.

It was at the mills, that we first appreciated the real luxury of sleeping in hammocks. One lays peacefully down without the annoying consciousness that he is beset with marauding, bloodthirsty enemies. Throughout the whole province of Pará, hammocks are universally used, and never, but on one occasion, while we were in the country, were we annoyed by flea or bug. The hammock is a pleasant lounge by day, as well as resting place by night, and the uncomfortable heat that might be felt in a bed, is entirely avoided. In the centre of the walls of rooms appropriated as sleeping apartments, are staples and rings, or suspension hooks, and the hammocks are swung across the corners. Sometimes, a post placed in the middle of the room, answers as a point of divergence, and thus, a great number of guests may be accommodated, in little space, and with no inconvenience.

There is one enemy, who, sometimes, approaches even a hammock, and takes a tribute from the unconscious sleeper, and that is the vampire bat. They are common enough any where, but about the mill, seem to have concentrated in disproportionate numbers. During the day, they are sleeping in the tiles of the roof, but no sooner has the declining sun unloosed the eve, than they may be seen issuing in long, black streams. Usually, we avoided all their intimacies by closing the shutters at sunset; but occasionally, some of them would find entrance through the tiles, and we went forth to battle them with all the doughty arms within our-reach, nor stopped the slaughter until every presumptuous intruder had bit the dust—or, less metaphorically, had sprawled upon the floor. Several thus captured, measured, each, upwards of two feet across the wings; but most were smaller. Of their fondness for human blood, and especially that particular portion which constitutes the *animus* of the great toe, from personal experience, I am unable to vouch; but every one in the country is confident of it, and a number of gentlemen, at different times, assured us, that they themselves had been phlebotomized in that member, nor knew of the operation, until a bloody hammock afforded indubitable evidence. They spoke of it as a slight affair, and, probably, the little blood that is extracted is rarely an injury. If the foot is covered, there is no danger, or if a light is kept burning in the room; and often, we have slept unharmed, thus guarded, where bats were flitting about, and squeaking, the night long. Cattle and horses are not so easily protected, and a wound once made, the bat returns to it every night, until proper precautions are taken, or the animal is killed by logs of blood.

In different parts of the mill, were the nests of a species of wasp, made of clay, and generally fastened upon the wall. But, several times, upon our boxes, books, or plants, they commenced their labors, constructing so neat a little edifice, that it was hard to consider them intruders.

Another incident was more home-like. Within the noisiest part of the building, and in an unused piece of machinery, a little house-wren had constructed her home, and would have reared her pretty brood, but, I am sorry to say, some egg-collecting stranger chanced that way.

One morning, we took the montaria, and started for Corientiores, a plantation, or rather, what once was a plantation, some three miles below. The sun was rising unclouded—the tide fell swiftly, and we skimmed, arrow-like, in our little craft, past leafy banks and flowery festoonings, and in a course more tortuous than that of a meadow brook. The kingfisher sat perched upon his overhanging branch, scarcely big enough to carry off the minnows he so intently watched for, and a jewel in the sunlight, with his back of golden green, and satin breast. Sandpipers flew, startled, across the stream, and the shrilly-cackling rail skulked away at our approach. A duck-hawk sat upon the summit of a leafless tree, fearlessly eyeing us. Huge fish leaped out of the water, in all the ecstasy of piscatorial bliss; and we drew from the general joyousness, good omens of a successful morning's work. Arrived at our destination, naught appeared but a house in the distance, almost concealed by shrubbery, and every where else, a tangled bush, with a few tall trees, from whose topes, numbers of large flycatchers were calling "Bentivee-Bentivee." Through this labyrinth, we toiled a couple of hours, shooting few birds, running heedlessly, and to our peril, into bees' nests, and leaving rags of clothes, and shreds of flesh, among the prickly sword-grass; until, at length, we were fain to give it up as a bad job, and, coming near the house, sat us down under the orange-trees, whose abundant fruit served somewhat to stay our longings for breakfast. A white man came to the door, and seemed disposed to be communicative; so we mustered our forlorn stock of Portuguese, and soon made considerable advances in his graces. He insisted upon our taking a cup of coffee, and, after a little more nodding and comprehending, an both sides, nothing would do, but we must add to coffee, fish and farinha; fresh fish, too, and of his own catching, and none the less agreeable, doubtless, for being presented us by his pretty wife. After breakfast, our friend sent out to the orange-tree, and soon brought us a brimming goblet of orangeade; and finally, before our departure, he had a number of breadfruits brought in, and the extracted seeds, much like chestnuts, roasted, with which he crammed our pockets. Verily, thought we, if this is the custom of the country, and the mere fact of one's being a stranger is a passport to such hospitality, and a sufficient apology for powder-smutted faces, and ragged garments, there is some little good left in the world yet. Here was this man, with so generous a heart, really one of the laziest squatters in the neighborhood, without a vestige of any sort of cultivation upon his premises, and, evidently enough, dependent for his support upon the fish he might catch in the stream: he would have felt offended, had we offered to pay for our entertainment, so we did what we could, by slipping some mementoes into the hand of a bright-eyed young Apollo, who was trotting about with the freedom of a wild colt.

The breadfruit tree, which we saw growing upon this place, sprang from a plant originally introduced into the Botanical Garden of Pará by the Government. A few of these trees are scattered over the province, but they are considered rather as ornamental than useful. In appearance, it is one of the most beautiful of trees; having a large, wide-spreading top, profusely hung with many-lobed leaves, nearly two feet in length, and of a bright green. The fruit is nearly spherical, six inches in diameter, green in color, and curiously warted upon tile surface. Within, it is yellowish, and fibrous, and contains a number of seeds, which are eaten roasted. There is a superior variety, that is seedless, and the whole of which is eaten.

Another common visiting place from the Mills was the Larangeira, or Orange Grove, a little settlement not far below Corientiores, where a lazy commandant mustered a few beggarly troops, for the security of thin part of the province. The most remarkable object here, was a cotton tree, measuring thirty-two feet in circumference, two feet above the ground. The height corresponded to this vastness, and we left it with a very lively impression of what Nature might do here, only give her the opportunity. Fortunately for settlers, her powers are somewhat restricted, and for one such monster, there are a hundred, little formidable, else were clearing the land out of the question. From the Larangeira, we received a variety of shells, the Helix pellis-serpentis, Anastoma globosa, Bulimus regius, and Helix comboides (Ferr.) One of the largest trees of the forest is the Masseranduba, or Cow tree, and, about Pará, they are exceedingly common. One, in particular, stands directly on the road, beyond the first bridge from the mill, and cutting into this, with our tresado, the milk issued at every pore. It much resembled cream in appearance and taste, and might be used as a substitute for milk in coffee; or, diluted with water, as a drink. It is, however, little used, except as a medicine, or for the adulteration of rubber. The wood of this tree is red, like mahogany, very durable, and used much for purposes where such timber is required. There are said to be eight varieties of trees known at Pará, and more or less common, which yield a milky sap. Other trees yield fragrant gums, and nearly or quite all these products are used for medicinal purposes.

At length, we prepared to leave the Mills, having enjoyed ourselves to the utmost in this our experience of Brazilian country life. We had seen every thing that we could have seen, and bad made a beautiful collection of birds and other objects. It was with regret that we bade adieu to Mr. Leavens, who had contributed so much to our comfort and pleasure. The sun had not risen, when, guns upon our shoulders, and accompanied by a black, with a basket for the carriage of any interesting plants, or other objects that we might desire to appropriate upon the road, we set forth. We passed several bridges, spanning little streams, and for ten miles, walked through the deep forest. The cries of monkeys resounded about us, and every now and then, there came a shrill sound, like that produced by whistling with the finger in the mouth. We frequently afterwards heard this same whistle, in different parts of the country, but never were able to ascertain from what it proceeded. Most likely a squirrel, but we were assured it was the note of a bird. We encountered a spider, leisurely crossing the road, that might rival the tarantula in bigness. A sharpened stick pinned him to the earth, and we bore him in triumph to town. Across his outstretched legs none of us could span, and his sharp teeth were like hawk's claws. This species spins, on web, but lives in hollow logs, and probably feeds upon huge insects, perhaps small animals, or birds. We collected specimens of a great variety of Ferns, Calandrias, Telanzias, and Maxillarias, and observed many rich flowers of which we know not the names. But we did recognize a Passion-flower, with its stars of crimson, as it wound around a small tree, and mingled its beauties with the overshading leaves.

Chapter VI

*START FOR CARIPÉ—ISLAND SCENE—ARRIVAL—
VICINITY—TOMB OF MR. GRAHAM—DINNER—
SHELLING IN THE BAY—VARIETIES OF SHELLS—
MARTINS—TERNS—NUTS AND FRUITS—MODE OF
FISHING—FOUR-EYED FISH—ANT TRACKS—
MOQUEENS—FOREST—CREEPING PLANTS—WILD
HOGS, OR PECCARIES—TRAPS—AGOUTIS—PACAS—
SQUIRRELS—BIRDS—CHAPEL AND SINGING OF THE
BLACKS—ANDIROBA OIL*

Our delightful visit at Magoary had incited a desire for further adventure, and ere a week had elapsed after our return, we were preparing to visit Caripé. Profiting by past experience, we secured a small canoe, having instead of a cabin, merely an arched covering towards the stern, denominated a tolda, and affording sufficient shelter for short voyages. This was manned by two stout negroes. Caripé is nearly opposite Pará, distant about thirty miles, but separated by many intervening islands. Among these, thirty miles may be a short distance or a very long one, as the tides favor; for there are so many cross currents running in every direction, that it requires great care to avoid being compelled to anchor, and lose much time. As to pulling against the tide, which rushes along with a six mile velocity, it is next to impossible.

We left Pará at midnight, two hours before low tide; and falling down about eight miles, received the advancing flood, which swiftly bore us on its bosom. There were two others of our party, besides A—— and myself; and one taking the helm, the rest of us stretched our toughening bodies upon the platform, under the tolda, determined to make a night of it.

Morning dawned, and we were winding in a narrow channel, among the loveliest islands that eye ever rested on. They sat upon the water like living things; their green drapery dipping beneath the surface, and entirety concealing the shore. Upon the mainland, we had seen huge forests, that much resembled those of the North magnified; but here, all was different, and our preconceptions of a forest in the tropics were more fully realized. Vast numbers of palms shot up their tall stems, and threw out their coronal beauties in a profusion of fantastic forms. Sometimes, the long leaves assumed the shape of a feather-encircling crest, at others, of an opened fan; now, long and broad, they drooped languidly in the sunlight, and again, like ribbon streamers they were floating upon every breath of air. Some of these palms were in blossom, the tall sprigs of yellow flowers conspicuous among the leaves; from others, depended masses of large fruits ripening in the sun, or attracting flocks of noisy parrots. At other spots, the palms had disappeared, and the dense foliage of the tree tops resembled piles of green. Along the shore, creeping vines so overran the whole, as to form an impervious hedge, concealing every thing within, and clustering with flowers. Very rarely, a tall reed was seen, and by the leaves which encircled every joint, and hung like tassels from its bended head, we recognized the bamboo. Frequently we passed plantations, generally of sugar cane, and looking, at a distance, like fields of waving corn; in beautiful contrast with the whole landscape beside. We lost the tide, and were obliged to creep along shore, for some distance, at the rate of about a mile an hour. At length, towards noon, turning a point, we opened at once into a vast expanse of water, upon the farther side of which the tree tops of Marajo were just visible. Immediately to our left, distant about a mile, and in a small circular bay, the broad white beach and glistening house upon its margin, told us we had arrived at Caripé. We were all enthusiasm with the beautiful spot, heightened doubtless by the approaching termination of our voyage; for in our cooped-up quarters, we were any thing but comfortable or satisfied. Moreover, a sail in the hot sun, unfortified by breakfast, tendeth not to good humor.

 Landing upon the beach, and having the canoe dragged up high and dry, we proceeded to the house, and soon made the acquaintance of the old negroes, who had charge of the premises. They set about preparing dinner, and we, meanwhile, slung our hammocks in the vacant apartments, and reconnoitered our position. The house was remarkably well constructed, for the country, covering a large area, with high and neatly plastered rooms, and all else conveniently arranged. In front was a fine view of the bay, and Marajo in the distance. Upon either side, the forest formed a hedge close by. Behind, was a space of a few acres, dotted with fruit trees of various kinds, and containing two or three thatched structures, used for various purposes; one of which particularly, was a kiln for mandioca. Here a black, shaggy goat, with horns a yard in length, lay enjoying himself in the drying pan. A number of young Scarlet Ibises were running tamely about. A flock of Troopials had draped a tree, near the house, with their nests, and were loudly chattering and scolding. But amid these beauties, was one object that inspired very different feelings. Close under our window, surrounded by a little wooden enclosure, and unmarked by any stone was the tomb of Mr. Graham, his wife, and child. He was an English naturalist, and with his family had spent a long time in the vicinity of Pará, laboring with all a naturalist's enthusiasm to make known to the world the treasures of the country. He left this beach, in a small montaria, to go to a large canoe, anchored at a little distance; and just as he had arrived, by some strange mishap, the little boat was overturned, and himself, his wife, and his child were buried beneath the surf. The bodies were recovered and deposited in this enclosure. Mr. Graham had been a manufacturer, and was a man of wealth. His family suffer his remains to lie moldering here, unmarked, although several years have elapsed since the catastrophe.

We were standing here, when a smiling wench announced dinner upon the table, and all reflections upon aught else were dissipated.

It is customary for persons visiting these solitary plantations to provide themselves with such provisions as they may want; but we were as yet uninitiated, and had secured nothing but a few bottles of oil and vinegar. But fish and farinha are the never failing resort, and to this we were now introduced with raging appetites. Here a slight difficulty occurred at the outset. The old woman had a store of dishes, but neither knife nor fork. We had penknives, but they were inconvenient, and tresados, but they were unwieldy; so, sending etiquette to the parlor, we took counsel of our fingers in this embarrassing emergency, and by their active co-operation, succeeded in disposing, individually, of a large platter of a well mixed compound, in which oil arid vinegar, onions, pepper and salt, materially assisted to disguise the flavor of the other two ingredients. There have been more costly meals, and perhaps, of a more miscellaneous character, than our first at Caripé; but I doubt if any were ever more enjoyed. After this dinner, we got on more genteelly, for we heard of a store in the neighborhood, and by as frequent visitations as our necessities rendered expedient, provided ourselves with every thing requisite. Fresh fish were abundant; and frequently some Indian in the vicinity would bring eggs, in exchange for powder and shot. Add to these a daily dish of mussels, or, more conchologically speaking, of Hyrias and Castalias, and our ways and means are explained.

We had come to Caripé more particularly for shells, inasmuch as it was the most celebrated locality for them in the vicinity of Pará. The bay so faces the channel, that the tides create a great surf and collect large numbers of various shells. We were just in time for the spring tides, when the water rises and falls fifteen feet; now, foaming almost to the top of the bank, now, leaving exposed a broad flat of sand, beyond which, in shallow water, is a muddy bottom. This latter was our shelling ground; and whenever the water would permit, all of our party, and the boatmen, were wading neck deep about the bay. Each carried a basket upon his arm, and upon feeling out the shell with his toes, either ducked to pick it up, or fished it out with scoop-nets made for the purpose. In a good morning's work we would, in this way, collect about one hundred and fifty shells. Those in the deeper water were of three varieties, the Hyria currugata (Sow.), the Hyria avicularis (Lam.), and the Anadonta esula (D'Orbigny), the latter of which was extremely uncommon. Nearer the shore, and in pools left standing in the sand, were the Castalia ambigua (Lam.), always discoverable by the long trails produced by their walking. Of three other small species we found single specimens, all hitherto undescribed by conchologists. Two of these were of the genus Cyrena, and the third an Anadonta. In the crevices of the uncovered rocks were great numbers of the Neritina zebra (Lam.), which variety is often seen in the market of Pará, and is eaten by the negroes. About one hundred yards east of the house, was a tide stream extending into the woods, and called in the country, *igaripé*. Here, and in similar igaripés in the neighborhood, were numbers of a red-lipped Ampullaria.

The water was so delightfully tempered, that we experienced no inconvenience from our long wadings, beyond blistered backs, and this we guarded against somewhat by wearing flannel. A kind of small fish, that bites disagreeably, was said to be common in these waters, and though we never met them, we thought it as well to encounter them, if at all, in drawers and stockings. The tide here fell with very great slowness; but at the instant of turning, it rushed in with a heavy swell, immediately flooding the flat, and breaking with loud roarings upon the shore. Besides the shells above enumerated, the Bulimus haemastoma was extremely common upon the land. Frequently we found their eggs. They were nearly an inch long, white, and within, was generally the fully formed snail, shell and all, awaiting his egress.

At low water, upon the bushes in some parts springing plentifully from the sand, large flocks of Martins, Hirundo purpurea, were congregated, like swallows in August. They seemed preparing for a migration, but as we saw them frequently throughout our journeyings, at different seasons, they probably remain and breed there. Flocks of Terns were skimming every morning along the beach, and as we shot one of their number, the others would fly circling about, screaming, and utterly regardless of danger.

The tides here collected great quantities of nuts and fruits, and along high-water mark, was a deep ridge of them, some, dried in the sun, others, throwing out their roots and clinging to the soil. We picked up an interesting variety of the palm fruits, and large beans of various sorts. One kind of the latter, in particular, was in profusion, and we soon discovered the tree whence they came, growing near by. It was tall and nobly branching, and overhung with long pods. Several varieties of acacias also ornamented the shore, conspicuous every where, from the dark rich green of their leaves. These, also, bore a bean in a broad pod, and the Indians asserted it a useful remedy for the colic. Here, also, we discovered a new fruit. It resembled much a strawberry, in shape, color, and flavor, except that its red skin was smooth, and its size that of a large plum. It covered in profusion the top of a large tree, and its appearance then was most beautiful. The negroes ate large quantities of it. We were told, afterwards, in the city, that it was a useful and agreeable medicine, having upon the system some of the beneficial effects of calomel.

Caripé is famous for its fishery, and we observed, with interest, the manner of taking fish in these igaripés. A matting is made of light reeds, six feet in length, and half an inch in diameter, fastened together by strings of grass. This, being rolled up, is easily transported upon the shoulder, to a convenient spot, either the entrance of a small igaripé, or some little bay, flooded by the tide. The mat net is set and properly secured, and the retiring tide leaves within it the unlucky fish. This mode is very simple, yet a montaria is frequently filled with the fish, mostly, of course, small in size. We saw a great many varieties thus daily taken, and much we regretted that our ignorance of ichthyology rendered it impossible for us to distinguish them, and that our want of facilities made it equally impossible to preserve them. One curious species, the Anableps tetrophlhalmus, was very common. It is called by the people, the four-eyed fish, and is always seen swimming with nose above the surface of the water, and propelling itself by sudden starts. The eye of this fish has two pupils, although but one crystalline and one vitreous humor, and but one retina. It is the popular belief, that as it swims, two of its eyes are adapted to the water, and two to the air.

It was curious to observe the tracks of the Saüba ants about the grass, in some parts near the house. By constant passing, they had worn roads two inches wide, arid one or more deep, crossing each other at every angle. These paths usually ran towards the beach, where quantities of food were daily deposited for the ants. A far greater nuisance than ants were Moqueens, little insects that live in the grass, and delight to attach themselves to any passerby. They are red in color, and so small, as to be scarcely distinguishable. But there is no mistaking their bite, and, for a little time, it produces an intolerable itching. We had known something of them at the Mills, but the dwellers there were nothing to those at Caripé.

The forest around us was mostly of second growth, and difficult of ingress, except along the road, which extended back, about two miles, to an old ruin. At this place, we noticed, in the doorway, a tree, nearly a foot in diameter, and yet, but a very few years had elapsed since the house was inhabited.

The creeping vines were of a different variety from any that we had before seen, contorted into strange shapes. One, particularly, with its broad stalk, resembled a shriveled bean pod.

Paths of wild hogs, or Peccaries, crossed the woods every where, these animals associating in droves. They much resemble the domestic hog, but never attain a large size. At various places, in these paths, were traps set by the negroes for Pacas and Agoutis, or other small animals. A thick hedge of limbs, and prickly palm leaves, is laid along, and any animal encountering this, will prefer following its course to making forcible passage, until his mortal career is probably terminated in a figure-four trap.

The Agoutis are small animals of the Rodentia family, of a reddish color, very common, and esteemed as food. They are much inferior in this respect, however, as well as in size, to the Pacas. These somewhat resemble Guinea pigs in form, and are the size of a young porker, living in burrows in the ground. They are very prettily spotted, and are a beautiful species.

In these woods, we saw a number of Squirrels, the same nimble things as squirrels elsewhere. There seems to be but one variety in the vicinity of the city, something smaller than our red squirrel, and of a color between red and gray. The place of this family is fully supplied by monkeys, which are seen and heard every where.

In the denser thicket we encountered a curious species of bird, which, afterwards, we found to be common throughout the province, in like situations. This was the White-bearded Puff-bird, Tamatia leucops. By collectors, at Pará, it is known by the name of Waxbill, from its long, red beak. This bird is the size of a jay, and almost wholly a lead color, approaching to black. It receives its name from the loose feathers upon the throat, which it has the habit of puffing out until its neck appears as large as its body. Owing to the secluded situations in which we found this bird, we could observe little of its habits, but another variety of the same family was common about the rice-mill, at Magoary, where, at any time, numbers of them might be seen sitting upon the top of some dead tree, whence they sallied out for insects, after the manner of the fly-catchers. They were very tame, and only learned caution after sad thinning of their numbers. This species is the Swallow Putt-bird, L. tenebrosa, and is nearly the size of a Martin. We discovered a nest of this bird. It was built in the fork of a limb, and both the nest itself, and the eggs which it contained, strikingly resembled those of our Wood-pewee, Muscicapa virens. A third small variety of this family is the Spotted, T. macularia, seen only in the deeper woods.

Connected with our house was a little chapel, upon the altar of which, was a rude representation of the Virgin, and every morning and evening, the blacks knelt in devotion. Upon certain evenings, all of them, and some of the neighbors, would come together, and, for an hour, chant the Portuguese hymn, in wild tones, but very pleasing. A lamp was constantly kept burning in this chapel. Similar customs prevail at most of the country sitios, and by many of the planters, the blacks are trained up rigidly to the performance of these observances.

The oil universally used for burning is obtained from the nuts of a tree known as the Andiroba. This tree is lofty and its widespreading top is overhung with large round pericarps, each of which contains eight nuts, of triangular shape. These are mashed between stones, and placed in the sun, which soon causes the oil to exude. It is dark in color, and burns with a dim light. Its taste is intensely bitter. It is considered a valuable remedy for wounds.

The torches used by the blacks, at Caripé, consisted merely of a few small nuts of a species of palm, strung upon a stick. They were full of oil, and burned clearly, answering their purpose admirably.

Chapter VII

LEAVE FOR TAÜAÜ—INDIANS—ARRIVAL AT MIDNIGHT—MORNING VIEW—THE ESTATE—TILARIA OR POTTERY—LIME KILN—SLAVES—CASTANHA TREE—CUYA OR GOURD TREE—ANT HILLS—AN ANT BATTLE—FOREST—MACAWS—DOVES—OTHER BIRDS—SLOTH—COATI——MACURA—BUTTERFLIES—RETURN TO THE CITY—FESTIVAL OF JUDAS—VISIT SR. ANGELICO, UPON THE GUAMÁ—BRAZILIAN COUNTRY HOUSE—CURIOUS AIR-PLANT—SERINGA OR RUBBER TREES—HARPY EAGLE—MONKEYS

Taüaü is one of the estates of Archibald Campbell, Esq., and by his invitation, we made arrangements for spending a few days there, in company with Mr. Norris. The distance from Pará is one tide, or about thirty miles, nearly south, and upon the river Acará. We left the city late in the afternoon, in the same canoe, and with the same boatmen, who accompanied us to Caripé. Just above the city, the Guamá flows in with a powerful current, setting far over towards the opposite islands. Passing this, we entered the stream formed by the united waters of the Mojú and Acará, and a few miles above, turned eastward, into the latter; a quit, narrow river, winding among comparatively lofty banks, and through large and well cultivated plantations. The clear moonlight added inexpressibly to the charm of this voyage, silvering the trees, and casting long shadows over the water. The blacks struck up a song, and the wild chorus floated through the air, startling the stillness. Frequently the same song came echoed back, and soon was heard the measured sound of paddles, as some night voyager like ourselves, was on his way to the city.

One cannot sail upon these streams, where unreclaimed nature still revels in freedom and beauty, without feeling powerfully the thickly clustering associations connected with them, and having often before his mind the scenes that have here transpired, since white men made this the theatre of their avarice and ambition. The great race who inhabited this part of the continent were the Tapuyas, whose name is now the general name for Indian. They were a kindly, hospitable race, the least cruel of all the Brazilian Indians, and received the whites with open arms. The whole main, and all these lovely islands, were their homes, and here, in peaceful security, they whiled away their lives like a summer's day. Henceforth their story is soon told. They were seized as slaves, mercilessly treated, their lives of no more value than the beasts of the wood. Countless numbers perished beneath their toil. Millions died from epidemic diseases, and many fled far into the interior, hoping to find some spot that the white man could never reach. The whole Tapuya race have disappeared, except here and there a solitary one, less fortunate, perhaps, than his nation.

As we approached Taüaü, the bank increased in height, and from some distance, the glistening tiles of a long building were conspicuous. At length, the large plantation house appeared upon the brow of the hill, almost concealed by the trees and shrubbery, and a light descending the steps betokened that our approach was observed. The overseer himself had come down to bib us welcome, and landing at the nicely sheltered wharf that projected into the stream, we followed him up the flight of stone steps to the house. A room in the upper story was ready to receive our hammocks, and here, we turned in, to await the morning. It was scarcely daybreak, when we were aroused by the entrance of a servant bringing coffee, and no further inducement was necessary to our early rising. The sky was unclouded, and the drops which had fallen during the latter part of the night, covered the trees with brilliants, as the sun broke upon them. Every thing smiled with the morning, the distant woods, the lake-like stream, the hill slope covered by orange and cocoa trees. Below, and a little to the right, was the tilaria, whose glistening roof had attracted us the night before, and numbers of blacks were already within, engaged at their work.

This estate was laid out by the Jesuits, and bears the marks of their good taste. The land, for a long distance from the river, is rolling, sometimes rising one hundred feet above the water level. The soil is of a fine red clay, and from this the estate derives its name, Taüaü signifying in the native tongue, red clay. Mr. Campbell is one of the largest manufacturers of pottery in the province. He labored hard to have fine earthen ware made, and was at expense in getting out a workman, and the requisite additional material. But the workman was unskillful, and the scheme, for the time, proved abortive, though probably practicable. The articles of ware most in demand, are water jars, and floor and roof tiles. The former are made upon the wheel, as elsewhere. The tiles are made by the women, floor tiles being about six inches square, by two thick, and roof tiles about fifteen inches long, six wide and one half inch thick, curved, longitudinally, into half a scroll. Near the house, was a kiln for burning lime. This was just finished, and being still unblackened by fire or smoke, was of singularly elegant appearance, with its dazzling white walls, and yellow moldings. The lime here burned is shell lime, and for this purpose, vast quantities of small shells are collected at Salinas, and other localities upon the sea shore. Upon the hill, and west of the house, stood a small chapel, and beyond this, extending a long distance upon the brow, were the houses of the blacks, structures made by plastering mud upon latticed frames of wood, and thatched with palm leaves. There were about eighty slaves connected with this plantation, some engaged in cultivating the ground, or laboring in the forest, others at the tilaria or the kiln. They were summoned to labor about five in the morning, by the bell, and were at work about two hours after dark; but during the heat of the day, they were allowed a long interval of rest. The chief overseer, or *fator*, was in the city, where, at this season, most whites throughout this vicinity were attending the festivals, but his place was supplied by a very intelligent mulatto. Upon Saturday afternoon, all the blacks collected around the store-room to receive their rations of fish and farinha, for the ensuing week. About twenty pounds of the latter was the allowance for an adult, and a proportionate quantity of fish; the whole expense averaging a fraction less than three cents per diem, for each person. Many of these blacks had fowls, and small cultivated patches, and from these sources, as well as from wood, and river, obtained much of their support.

 Beyond the tilaria, was a long swamp, and here, a number of jacanas, snipes, and plovers, were constantly flying about, and screaming their call notes. Back of the house, was a grove of fine trees, some, apparently, having been planted for ornament, others, bearing profusion of various sorts of fruits. The one of all these most attractive, was that which produces the Brazil nut, called in the country castanhas. Botanically, it is the Bertholletia excelsa. This tree was upwards of one hundred feet in height, and between two and three in diameter. From the branches were depending the fruits, large as cocoa-nuts. The shell of these is nearly half an inch in thickness, and contains the triangular nuts, so nicely packed, that, once removed, no skill can replace them. It is no easy matter to break this tough covering, requiring some instrument, and the exercise of considerable strength: yet we were assured by an intelligent friend, at the Barra of the Rio Negro, that the Guaribas, or Howling Monkeys, are in the habit of breaking them, by striking them upon stones, or the limbs of iron-like trees. This friend related an amusing incident of which he had been witness, where the monkey, forgetful of every thing else, pounding down the nut with might and main, in a fever of excitement, struck it with tremendous force upon the tip of his tail. Down dropped the nut, and away flew monkey, bounding and howling fearfully. How long the victim was laid up by his lame tail, our friend was unable to inform us; but we thought one thing certain, that monkeys had changed since Goldsmith's day, inasmuch as, at that time, as we are informed, the tip of a monkey's tail was so remote from the centre of circulation as to be destitute of feeling. When the castanha nuts are fresh, they much resemble, in taste, the cocoa-nut, and the white milk, easily expressed, is no bad substitute for milk in coffee. This soon becomes rancid, and at length turns to oil. The nuts are exported largely from Pará, and are said to form a very important ingredient in the manufacture of sperm candles.

There is another nut, probably of the pot tree, Lecythis ollaria, mentioned by Spix, much resembling the castanha in appearance and growth. When this is ripe, an operculum falls from the lower side of the encasing pericarp, and affords egress to the nuts within. Monkeys and squirrels are so excessively fond of these, that it is usually impossible to obtain more than the empty pericarp.

Next to the castanha tree, the calabash, or cuya, was most attractive. It was low, its trunk overgrown with moss and small parasitic plants. Directly from the bark of the trunk, or branches, without intervening stems, grew the gourds, a bright green in color, and often six inches in diameter, giving the tree a very curious appearance. The smaller gourds are cut in halves, the pulp removed, and the shell reduced by scraping. This, being sufficiently dried, is painted both inside and out, by the Indian women, with ingenious and sometimes beautiful devices. They are the universal drinking cup, and are known by the name of *cuyas*.

The cleared space, round about, was of great extent, much being under cultivation, but a still larger portion was thickly overgrown with tall weeds. Here were scores of ant hills, between three and four feet in height, conically shaped, and each having two or more entrances the bigness of one's arm. The exterior of these hills was of stony hardness; within were galleries and cells. The earth of which they were composed seemed always different from that in the vicinity, and evidently had been brought grain by grain. In the woods, we frequently encountered a different kind of ant hill. A space of a rod square would be entirely divested of tree or bush, and every where, the surface was broken into little mounds, formed by the earth brought up from below. While upon this subject, I will describe an ant battle, several of which we watched, at different times and places. The combatants were always a species of small black ant, and a red variety, equally small. Coming in long lines from different directions, it seemed as if they had previously passed a challenge, and had selected the ground for their deadly strife. The front ranks met and grappled, toiling like wrestlers, biting and stinging; they soon fell, exhausted and in the death agony. Others fought over their bodies and likewise fell, and still, continually, over the increasing pile, poured on the legions of survivors, fighting, for several days in succession, until a pile of a peck, or more, lay like a pyramid. They marched to certain death, and had their size been proportionate to their courage, these battle fields had mocked earth's bloodiest.

The woods about Taüaü were of the loftiest growth and filled with game, both birds and animals. Here we first encountered the gorgeous Macaws, climbing over the fruit-covered branches and hoarsely crying. They were wiser than most birds, however, having acquired something of that faculty from long experience; for their brilliant colors, and long plumes render them desirable in the eyes of every Indian. They were not unwilling to allow us one glimpse, but beyond that, we never attained.

As might be expected, Woodpeckers are exceedingly numerous throughout these forests, and the size of most species is in some proportion to the labor they have to perform, in gaining their livelihood from these enormous trees. Every where is heard their loud rattle, and harsh, peculiar note. In this latter respect, many species so resembled those familiar to us at home, that we could scarcely believe that the stranger that fell dead at our feet, victim of a long, successful shot, ought not to have been one of the Golden-wings, or Red-heads, that we had so often tried our skill upon.

The same varieties are found throughout the river country; as common upon the Rio Negro as at Pará. The most gaudy of all, and the especial favorite of the Indians, is the Picus rubricollis, whose crested head, neck, and breast, are of brilliant red. Another finely crested species is the P. lineatus. There is also the P. fulvus, nearly the size of our Golden-wing, and of a deep brown color. Another, as large, is almost wholly of a light yellow. Of lesser species, there seemed no end, and some of them were singularly diminutive.

The Tree-creepers were a more eagerly sought family, and two beautiful little species are quite common in the vicinity of Pará. One of these is of a deep indigo blue, with a blank throat, Certhia coerulea; the other, C. Cayana, is conspicuous for the brilliant ultramarine blue that caps his head. Otherwise he is marked with blue, and black, and yellow. These little things are usually seen running up and down the tree trunks, or flitting hurriedly from branch to branch, busied in searching for insects upon the bark. They are extremely familiar, and allow of near approach. At intervals, they emit slight, whispering notes, but their anxious haste leaves one with the impression that they might do themselves much more credit as songsters, at their leisure. We never fell in with these species up the river, their place there being supplied by other varieties.

In the lower woods, were great numbers of Doves, of many species, but similar to those we had elsewhere met. Most beautiful of all is the Pombo troucal—Columba speciosa (Linn.), the "bird of the painted breast." They are of large size, and usually are seen in pairs, within the shade of some dense tree, but, early in the morning, are often discovered, in large numbers, upon the limbs of leafless trees, of which, at every season, there are very many throughout the forest.

The smallest and most graceful of all these doves, is the Rola, the Ground Dove—C. passerina—of our Southern States, not larger in size than many sparrows. They are seen, about cleared fields and houses, in large flocks, and when unmolested, become extremely familiar.

About every plantation, are two varieties of Tanagers, domestic as our robin, resting in the orange trees under the windows, and constantly flitting among the branches, uttering their few notes, which, though pleasing, can scarcely be called a song. One of these, the Silver-bill, Tanagra jacapa, has a crimson-velvet livery, and silvery bill; the other, Tanagra cana, is, mostly, a sky-blue. The former is called Pipira, from its note. Its nest is neatly formed of leaves and tendrils of vines, and the eggs are usually three and four of a light-blue color, and much marked, at the larger end, with spots of brown.

Upon one occasion, A—— brought in a sloth which he had shot, and I skinned him, with the intention of preserving his body for some anatomical friend, at home, to whom sloths might be a novelty. But our cook was too alert for us, and before we were aware, she had him from the peg where he hung dripping, and into the stew-pan, whence he made his debut upon our dinner-table. We dissembled our disappointment, and did our best to look with favor upon the beast, but his lean and tough flesh, nevertheless, could not compare with monkey.

There are animals much resembling the raccoon, called Coatis. They are extremely playful, and may occasionally be seen gamboling, in parties of two or more, among the dry leaves. When tame, they possess all a raccoon's mischievousness. These, as well as monkeys, according to Goldsmith, were wont, of old, to live upon their own tails.

One of the negroes brought us a little animal of the opossum kind, called the Macura cheçhéga. It was scarcely larger than a small squirrel, and its hair was of silky softness. We could probably have preserved it alive, but its captor had broken both its hinder legs, to prevent its running away. This is the common custom of the blacks and Indians, when they desire to preserve an animal for a time, before it is eaten.

About the flowers in wood and field, was a profusion of butterflies, almost all gaudy beyond any thing we have at the North. The most showy of all, was a large variety, of a sky-blue color, and brilliant metallic lustre. We observed but one species seen also in the Northern States, the common red butterfly of our meadows, in August. In this clime, the insects of all kinds are nimble, beyond comparison with those elsewhere, and often, the collector is disappointed in his chase. He has a more embarrassing difficulty than that, however, for without the most unceasing care, the ever-present ants will, in a few moments, destroy the labor of a month.

A week passed rapidly and delightfully. The fator returned, and urgently pressed our longer stay, but reported letters from home, hastened us back to the city. The past week had been the close of Lent, and during our absence, the city had been alive with rejoicings. Festas and celebrations had taken place daily, and hundreds of proprietors, with their families and servants, had collected, from every part, to share the general joyousness. Of all these festival days, that of Judas was the favorite, and the one especially devoted to uproariousness. That unlucky disciple, by every sort of penance, atoned for the deeds done in the flesh. He was drowned, he was burned, he was hung in chains, and quartered, and was dragged by the neck over the rough pavements, amid the execrations of the rabble.

A few days after our return from Taüaü, in company with Messrs. Smith and Norris, we visited the plantation of Sr. Angelico, upon the river Guamá, for the purpose of seeing the manufacture of rubber. A few hours' pull brought us, by sunrise, to a sitio upon the southern side, standing upon a lofty bank, and commanding a fine view of the river. Here we exchanged our canoe for a montaria, as we were soon to ascend a narrow igaripé, where a few inches of width, more or less, might be material; after which, we continued a little distance further up the river. The Guamá is a larger stream than the Acará, but much like that river in the appearance of its banks, these often being nigh, and, in parts, well settled. By some of the eastern branches of the Guamá, easy communication is had with streams flowing towards Maranham, and this route is occasionally taken by carriers. Suddenly the boat turned, and we shot into a little igaripé so embowered in the trees, that we might have passed, unsuspecting its existence. The water was at its height, calm as a lake. Threading our narrow path between the immense tree trunks, a dozen times, we seemed to have reached the terminus, brought up by the opposing bank; but as often, a turn would discover itself, and we appeared as far from the end as ever. Standing in this water were many seringa, or rubber trees, their light-gray bark all scarred by former wounds. We gave passing cuts at some of them, and saw the white gum trickle down. When, at last, we landed, it was to pick our way, as best we could, over a precarious footing of logs and broken boards, from which a false step might have precipitated us into mud, rich and deep. Once upon terra firma, a short walk brought us to the house, concealed among an orchard of cocoa trees. A loud *viva* announced our approach, and immediately, Senhor Angelico bustled out of his hammock, where he lay swinging in the verandah, and, in his night-gown, bade us welcome. He was a confidence-inspiring old gentleman, with his short, stout body, and twinkling eyes, and a chuckling laugh that kept his fat sides in perpetual motion, belying, somewhat, his tell-tale gray hairs, and his high-sounding title of Justicia de Paz.

The Senhor did not forget the necessities of early travelers. A little black boy brought around fresh water for washing, and, in a trice, breakfast was smoking on the table, our host doing the honors with beaming face, and night-gown doffed.

This was the first decidedly Brazilian country house that we had visited, and a description of it may not be uninteresting. It was of one story, covering a large area, and distinguished, in front, by a deep verandah. The frame of the house, was of upright beams, crossed by small poles, well fastened together by withes of sepaw. A thick coat of clay, entirely covered this, both within and without, hardened by exposure into stone. The floors were of the same hard material, and in front of the hammocks, were spread broad reed mats, answering well the purpose of carpets. Few and small windows were necessary, as the inmates of the house passed most of the day in the open air, or in the verandah, where hammocks were suspended for lounging, or for the daily siesta. The roof was of palm thatch, beautifully made, like basket work in neatness, and enduring for years. The dining table stood in the back verandah, and long benches were placed by its sides, as seats. Back of the house, and entirely distinct, was a covered shed, used for the kitchen and other purposes. Any number of little negroes, of all ages and sizes, and all naked, were running about, clustering around the table as we ate, watching every motion with eyes expressive of fun and frolic, and as comfortably at home as could well be imagined. Pigs, dogs, chickens and ducks assumed the same privilege, notwithstanding the zealous efforts of one little ebony, who seemed to have them in his especial charge. Do his best, he could not clear them all out from under the table at the same time. They knew their rights. But these little inconveniences one soon becomes accustomed to, and regards them as matters of course. The house stood in a grove, and round about, for some distance, what had been a cultivated plantation was growing up to forest, the Senhor having turned his attention to the seringa. Scattered here and there, were neat looking houses or the blacks, many of whom were about, and all as fat and happy as their master. It was amusing to see the little fellows, crammed full of farinha, and up to any mischief, come capering about the Senhor, evidently considering him the best playmate on the premises. He enjoyed their frolics exceedingly, and with a word or a motion would set them wild with glee. It is this universally kind relation between master and slaves in Brazil, that robs slavery of its horrors, and changes it into a system of mutual dependence and good will.

We strolled about the woods several hours, shooting birds and squirrels, or collecting plants. Some of the air plants found here produced flowers of more exquisite beauty than we ever met elsewhere, particularly, a variety of Stanhopea, which bore a large, while, bell-shaped flower. This we succeeded in transporting to New York, and it is now in the green-house of Mr. Hogg, together with many other plants of our collecting. Under his care, they promise to renew the beauty of their native woods. We engaged a score of little hands to pick up the shells of the B. haemastoma, which in some places strewed the ground. Why so many empty shells were there, it was impossible to understand. The Senhor asserted that the animals vacated their shells yearly. A—— shot an armadillo in the path, which was served up for our dinner. The flesh resembled, in appearance and taste, young pork.

In the afternoon, rain commenced pouring, and we were obliged to take to our hammocks in the verandah, amusing ourselves as we might. All night long, the rain continued, and to such a degree, that it was found impossible to collect the sap of the seringa. Greatly to our disappointment, therefore, we were obliged to return ungratified in the main object of our visit, although in every other sense, we had been richly repaid. We had, afterwards, opportunities of observing the manufacture of shoes, which, in its proper place, will be described. Why rubber should be designated by the barbarous name of caoutchouc, I cannot tell. Throughout the province of Pará, its home, it is universally called seringa, a far more elegant, and pronounceable appellation, certainly.

On our way down the river, we saw the nose of an alligator protruding from the water, as he swam up the current. These animals very rarely are met in these streams, and indeed, throughout the whole lower Amazon region, excepting in the islands at the mouth of the river, where they abound.

While absent upon this excursion, Mr. Bradley, an Irishman, who trades upon the upper Amazon, arrived at Mr. Norrie's, bringing many singular birds and curiosities of various kinds. One of the former, was a young Harpy Eagle, a most ferocious looking character, with a harpy's crest, and a beak and talons in correspondence. He was turned loose into the garden, and before long, gave us a sample of his powers. With erected crest, and flashing eyes, uttering a frightful shriek, he pounced upon a young ibis, and, quicker than thought, had torn his reeking liver from his body. The whole animal world below there, was wild with fear. The monkeys scudded to a hiding-place, and parrots, herons, ibises, and mutuns, with all the hen tribe that could muster the requisite feathers, sprang helter-skelter over the fences, some of them never to be reclaimed.

A less formidable venture was a white monkey, pretty nearly equal, in his master's estimation, to most children, and some adults. Nick had not been with us long, before he was upon the top of the house, and refused all solicitations to come down. It was of no use to pursue him. Moving slowly off, as though he appreciated the joke, he would at last perch upon some inaccessible point, and to the moving entreaties of his master, would reply by the applied thumb to nose, and the monkey jabber of "no, you don't." At other times, when there was no danger of sudden surprises, he amused his leisure by running over all the roofs in the block, raising the tiles, and peering down into the chambers, to the general dismay. At length, as fair means would not do, foul must; and Nick received a discharge from a gun loaded with corn. But somewhere upon the roof, he obtained a rag of cloth, and holding it before him, he would peep over the top, ready to dodge the flash. It would not do; we gave Nick up as lost; but of his own accord, he, at last, descended, and submitted to durance.

Chapter VIII

Leave Pará for Vigia—Boatmen—Inland Passage—Egrets and Herons—Stop at Sugar Plantation—Cupuassu—Mangroves—Insolence of Pilot—Vigia—Arrival at Sr. Godinho's—Reception—The Campinha and its Scenery—Sporting—Parrots—Employees—Sun-bird—Boat-bill—Tinami—Iguana Lizard—Sugar Cane—Mill—Slaves—Leave the Campinha—Kingfishers—Go below for Ibises—Sand-flies—Return to Pará—A Pet Animal

Soon after Mr. Bradley's arrival, Dr. Costa, the chief judge of the district of the Rio Negro, also arrived in Pará, upon his way to Rio Janeiro, and learning that we desired to visit the towns upon the Amazon, very kindly offered us his galliota and Indians for that purpose. So tempting an offer allowed of no hesitation, but as Mr. Bradley was to be in readiness to make the same journey, in a few days, we determined to await his convenience, and meanwhile, to make a short excursion to Vigia. This town is about fifty miles below, near the junction of a small tide stream with the Grand Pará. As the direct passage down the river offered little of interest, and, moreover, at this still squally season, was somewhat hazardous in a small canoe, we determined on the inland course, winding about among the islands, and requiring, perhaps, double the time.

We left Pará, on the first of May, in the same canoe that carried us to Magoary, and with the same negroes whom we had heretofore employed. These fellows, by long acquaintance, assisted by a modicum of their own good nature, and a due sense of our generosity, had molded themselves pretty much to our wishes. Unmerited oblivion ought not yet to overtake these good companions of our wanderings, and who knows but that a charcoal sketch of their lineaments and characteristics, may discover them to the notice of some other travelers, who may hereafter have like necessities with ourselves. And first, our round-faced, jolly-looking, well-conditioned Faustino; somewhat less a beauty, perhaps, than Nature intended, by reason of undisguiseable tracings of small-pox. Yet many a worse failing might be amply redeemed by the happy smile that ever lightened up his coal-black countenance, particularly, when enlivened by the slightest possible infusion of cashaça, which, as with the Rev. Mr. Stiggins, is his weakness. Faustino is a famous story-teller, and enacts his own heroes with a dramatic effect that is often very amusing. He is gifted in song too; and many a night, have his sweet catches softened our hard couch, and hushed us to sleep.

Faustino's companero, doubtless, once claimed a name proper; but long since, it seems to have been absorbed by the more distinguishing and emphatic designation of Checo, which in this country, signifies "small," a name by no means inapt. A Greek proverb says, "there is grace in the small;" but Checo has been a soldier, and now Checo's right eye is cocked for the enemy, and his left has an expressive squint toward the remote thicket. Nor do his eyes belie him, doubtless, for though he can wear out the night with his adventures in the southern provinces, no scar disfigures his anteriors or posteriors, as he sits glistening in the sun, naked as the day he was born. But Checo is faithful, and abhors Cashaça.

Besides these two, we were forced to take a pilot, on account of the intricacy of the passage, and, therefore, a lazy, villainous-looking mixture of Brazilian and Indian, sat at the helm; while a boy, like a monkey, whom he brought on board for what he could steal, was annoying us perpetually.

As there were no occupants of the cabin but A—— and myself, we had a comfortable allowance of room, wherein to stretch ourselves: and about us, in ship-shape order, upon the cabin sides, were piled our baggage, implements, and provisions; among which latter, farinha, bread, and molasses predominated. Knives and forks, spoons and plates, completed the furniture of our cuisine; and our table-cloth was a Turkish rug, whose more legitimate office it was to "feather our nests" at night.

Before dark, we had left the river, and starlight found us ascending a stream, in no wise distinguished in the character of its scenery from those which I have heretofore described; and yet, perpetually interesting from the ever new views that constant windings presented, and which required neither sunlight nor moonlight to cause us to appreciate their loveliness. With the changing tide, we anchored, and turned in for the night. It was amusing always to observe with what indifference our boatmen would stretch themselves out upon the seats, unprotected, in any way, from rain or dew, and drop at once into a profound sleep, ready, at an instant's warning, to start again to the oars. The pilot had brought along a hammock, which he swung between the masts, high above the others' heads; thus obtaining a situation that might have been envied by his masters, had not frequent acquaintance with hard resting-places somewhat weakened their sensibilities.

Some hours before daybreak, we were again under way; and the first glimpse of light found us exchanging the cabin for the deck, where, guns in hand, we planted ourselves, ready to take advantage of any unsuspicious Egrets that might be feeding upon the muddy bank. These Egrets, or Carças, as they term them in Brazil, are small, and of a snowy white, the Ardea candidissima; and are a very interesting addition to the river beauties, as they stalk along the banks, or sit perched upon the bushes, in the distance, resembling so many flowers. The stream was narrow, and the canoe was steered to one side or the other, as we saw these birds; and thus, until by repeated alarms, and much thinning of their ranks, they had become shy of our approach, they afforded us constant sport. Sometimes, far in the distance the keen eyes of the men would descry the Great Blue Heron, the Ardea herodias; and with silent oars and beating hearts, we crept along the shore, hoping to take him unawares. But it was of no avail. His quick ear detected the approaching danger; and long before we could attain shooting distance, he had slowly raised himself, and flown further on, only to excite us still more in his pursuit.

About nine o'clock, we stopped at a small sugar estate, where we proposed to remain over the tide. In landing, I inadvertently stepped off the blind stepping-stones, and brought up all standing with my knees in the mud, and slippers almost beyond redemption. However, I contrived to hook these out, and marched, in stocking feet, the remainder of the distance to the house, presenting, doubtless, an appearance as diverting as pitiful. But the whites and negroes who crowded the verandah, and awaited our approach, seemed too much accustomed to such mishaps to mind them, and a quickly applied liniment of agua fresca soon put all to rights again. We strolled into the woods, and after chasing about until we were weary, returned with several birds, mostly motmots and doves, and a number of the fruits called cupuassu. These are of the size and shape of a cocoa-nut in the husk, and within the shell is a fibrous, acid pulp, of which a delightful drink is made, much like lemonade. The producing tree is common in the forest, and of great size and beauty. The afternoon was rainy, and we were confined below. But the time passed not at all tediously, for beside the preserving of the birds, we had store of books wherewith to beguile our leisure. Next morning, we shot some rail, skulking among the mangrove roots by the water's edge. These birds are called from their notes Cyracúras, and are heard upon all these streams, in the early morning, or the dusk of evening, loudly cackling. It is unusual to observe more than one in a place, but at considerable distances, they call and answer each other. This is one of the birds that the citizens delight to domesticate. We heard also the sharp, quickly repeated notes of the Sun-bird, the Ardea helias, and the most beautiful of the heron tribe. Almost every bird is named in this part of Brazil, from its note, but this, by way of distinction, is called the pavon, or peacock. These birds were shy and we yet were ungratified by seeing one.

The mangroves that skirt all these streams are a curious feature. The tree itself is low, and has a small stem; but from this, radiate in every direction towards the water, long, finger-like branches. These take root in the mud, and are really the roots of the tree, supporting the stem at some distance above the water. When they are small, they serve for arrows to the Indians, being very light, and often perfectly straight. They not only so bind the soil as to prevent its wearing away by the constant flowings of the tide, but catch all sorts of drift, which in this way, contributes to the body of the island. Indeed, whole islands are thus formed; and within the memory of residents, an island of considerable size has sprung up within sight of the city of Pará. In a similar way, the thousands of islands that dot the whole Amazon have been formed.

Ever since we left Pará, our pilot had been inclined to insolence, but this afternoon, from the effects of cashaça which he had obtained at some of our landings, became intolerable. A——, at last, took his jug from him and pitched it overboard, giving him to understand that its owner would speedily follow, unless he changed his tone. This cowed the fellow into better manners, and A—— sent him forward, taking the helm himself. No traveler will care to employ a second time one of these low whites or half breeds.

Towards evening, as we approached Vigia, we came upon a bank, where a large flock of Garças, mixed with Herons, Spoonbills, and Scarlet Ibises were feeding. This was the first time we had seen the latter, but the sun was too low to discover all their beauty. By eight o'clock we had anchored off Vigia. This town had once been populous, and even contained a Jesuit college; but long since, the houses had gone to decay, and the forest encroached upon the streets. It is now principally inhabited by fishermen, and in the distant view, appears like Pará, the same building material being used. We were not to stop here, as our letters were to Senhor Godinho, who lived upon a small igaripé opposite the town, distant a few miles; therefore we were early under way, although the tide was against us. In a high bank which we passed, were several holes of Kingfishers, and numbers of the birds, some very small, others, twice the size of our Kingfisher of the North, were flying about. At length, we turned into the desired igaripé, and by dint of hard rowing and poling, advanced as far as the shell of a house stack upon the bank, whither our pilot went for directions. The fellow kept us waiting a half hour, and we pushed off without him, pleased enough to repay his villainies by a long walk through the mud and bushes. But the tide was out, and we lodged immovably in the mud, and for an hour's space, were fain to keep ourselves in as good humor as we might under a burning sun, until the tide came to our relief. A beautiful red hawk sat near by, eyeing our movements, and a flock of buzzards were eating the crabs along the exposed mud. Numbers of little Sandpipers, the Totanus solitarius, were running about, hasting to get their breakfasts before the flooding waters should return. There were many dead fish lying about, often of large size. We afterwards learned that these had been killed by poison thrown into the holes which they frequent at low water.

As the tide rose, we pushed slowly on, and soon opened into a large clear space, at the remote end of which appeared the plantation house. Senhor Godinho met no upon the dock which ran directly by the side of his mill, and welcomed us in good English with the greatest warmth and politeness. We at once, felt ourselves at home. Forthwith, our luggage was unstored, a room was opened to the light, very much to the astonishment of the bats and cockroaches, and the blacksmith made his appearance with hooks and staples for our hammocks. We followed the Senhor to the verandah above, and under the cool breeze, soon lost all thoughts of our morning's broiling. Every thing about indicated opulence and plenty. Blacksmiths, carpenters and masons were at work in their different vocations; the negroes and oxen were driving the sugar mills; the steam pipe of the distillery was in full blast; and stacks of demijohns and jars were piled in the rooms, or standing ready to receive the cashaça or molasses.

The house was surrounded by woods, some nearer, some farther; and directly in front of the verandah, was an intervening swamp, along whose edges, cyracuras were feeding, and in the middle of which, pigs and goats disputed empire with various small water birds, and a tame white heron. Beyond, to the left, and extending several miles, was a prairie or *campo*, crossed by parallel strips of woods, and the loud cries of parrots and toucans came swelling on the breeze. This was irresistible, and as soon as we could dispatch a hearty dinner, guns in hand, we sallied no a tour of exploration. The trees were all low, and the ground was crossed in every direction by the paths of the hogs, who roamed over these campos, half tamed, in immense numbers. Water lay upon the surface of the ground, often to considerable depth, but that we little cared for. We soon discovered the palms upon which the parrots were feeding, and in a short time, the boy who accompanied us, was loaded with as many of these birds as he could carry. The large parrots, as they fly slowly along, have a very conjugal appearance; always moving in pairs, side by side, and each and all discoursing with a noisy volubility that must destroy the effect of what they have to say. When one from a pair is brought down, it is amusing to see the survivor continue chattering on, without missing a word, or altering his course; altogether exhibiting a cool self-possession most anti-conjugal. Returning to the house, we busied ourselves in preserving such specimens as we wanted, the Senhor looking on with great interest, and relating anecdotes and histories of different animals and birds thereabout, and which, in his solitude, he had both time and inclination for observing. In the morning, we were out again, and indeed, were thus occupied every morning for a week, constantly obtaining something new and curious, besides keeping the table well supplied with game. It seems as heterodox to eat parrot as monkey, yet fricasseed parrot might rank favorably with most kinds of wild game. In a day or two, one of the Senhor's men, a free mulatto, six feet in height, straight as an arrow, and with an eye like a hawk, was enlisted in our service, through his master's kindness. Gregorio had a companero, an Indian, of like characteristics and propensities, called Francisco, and between the two, we were under a press of business. One of the birds which they procured for us, was the much desired Sun-bird. it was small, and exquisitely marked, "its plumage being shaded in bands and lines with brown, fawn color, red, gray and black, recalling to our minds the most beautiful of the nocturnal Lepidoptera." We frequently saw this bird domesticated in other parts of the province, and in this state it becomes exceedingly familiar, living entirely on flies and other insects. Another species as curious as the last, though not for its beauty, was the Boat-bill, Cancroma cochlearia. It is of the heron kind, but unlike its congeners, each mandible is shaped like half a keeled boat, short and broad. From the head, long plumes extend far down the black. One would think that nature delighted to give the most fantastic shapes to her handiwork in these climes. Besides these dwellers of the water, were Herons of various sorts, Snowy White, Blue, *et alii*, in profusion. The woods afforded us most of the species we had observed elsewhere, and many others entirely new. Here, a singular family was the Tinamus, gallinaccous birds, resembling pheasants in their habits, but shaped more like rails than any other bird, having long, slender necks, and scarcely any tails. They are universally known by the name of Inambu, and different species of the family are found throughout northern Brazil. The eggs of these birds are of the deepest green, and are superior to those of domestic fowls in taste. Here also were large, reddish-brown, Cuckoos, moving stealthily about the low trees, uttering, at intervals, the note, which, so generally, characterizes the family, and searching for caterpillars, and, it may be, the eggs of the little and defenseless birds. The common species is the Cuculus Cayanus, rather larger than our Yellow-billed Cuckoo, but of inferior beauty. Another species much resembling this in color, but of half the size, is often seen, and with far greater familiarity than the Cayanus, comes into the orange and cuya trees, about the houses, in search of worms' nests.

Upon the campo, where flocks of Red-breasted Orioles, Icterus militaris, of a deep brown color, except upon the breast and throat, which glow with a rich red. These birds have rather the habits of starlings than orioles, being usually seen upon the ground, or upon the low bushes, which, here and there, diversify the campo.

Here was also a large variety of Lapwing, called Terra-terra, from its loud and constantly repeated note.

By the brooks, which crossed the paths through the trees, numbers of pretty Doves, of all sizes, were cingregated, now, proudly strutting with outspread tails and drooping wings, now, chasing each other about the sandy margin, and now, with ruffled feathers, bathing themselves in the limpid water, and tossing the cooling drops over their shoulders.

Among the low shrubs, and about the cocoa trees, near the house, were many small species of birds, none prettier than the Tingtings, Tanagra violacea and T. chlorotica, two species of small Tanagers with shell-blue backs, and yellow breasts, frequently seen in cages in Pará. There was one other cage bird we sometimes met, called the Rossignol, or Nightingale, neither more nor less than a yellow-shouldered black oriole. It sings well, but scarcely deserves its honored name.

Besides the birds, we had a constant supply of monkeys and other animals for the table. Our pilot labored zealously to reinstate himself in our good graces, and brought in various articles which he thought would assist him in effecting his purpose. One of his captures was a live Iguana, called, in Brazil, a Chameleon, a lizard of four feet length. He had shaken the beast from a tree, upon the leaves of which it was feeding, and seizing it by the neck and the small of the back, made it his prize. This fellow was of a greenish color, and spotted. Upon his back were spines which he could erect at pleasure. Upon the ground, the iguanas move slowly, and their tail is then a powerful defensive weapon against their enemies, capable of inflicting a terrible lash, as this specimen showed us, after its arrival in the city. They are much esteemed as food, and their eggs are sought after with avidity for the same purpose. Although their food consists mostly of leaves and fruits, yet they rob the nests of birds, as do other lizards.

Senhor Godinho was one of the most extensive planters of the province, and interested us greatly by his agricultural and other information. The cane used in his mills was grown upon the borders of the igaripé, in different localities; and so inexhaustible is this rich alluvium, that it requires replanting but once in from sixteen to twenty years. Two mills constantly employed were insufficient to dispose of his yearly crop, and a large outhouse was filled with cane, half ruined in consequence. Most of the syrup was converted into cashaça, that being considered more profitable than sugar or molasses. Instead of tuns for the liquor in the distillery, hollowed tree

trunks were used, one alone of which contained twenty-five pipes' bulk. In the troubles of '35, the Senhor was compelled to flee the country, as were all other planters who could, and in the sacking of his place, sustained great loss. He was a self-made Portuguese, formerly a merchant in Pará, and his ideas were more liberal than those of his countrymen generally, as was evident enough, from his adoption of improved machinery for the manufacture of his sugar, instead of the methods in use at the time of the conquest. There were about one hundred slaves employed upon the plantation, and they seemed to look up to the Senhor with a pride and affection, which he fully reciprocated. He told us that, for months together, he was not obliged to punish one of them. They all had ways of earning money for themselves, and upon holidays or other times, received regular wages for their extra labor. There was a novel custom here, usual upon these retired plantations. Soon after sunset, all the house servants, and the children of the estate, came in form to ask the Senhor's blessing, which was bestowed by the motion of the cross, and some little phrase, as "adios."

It was with regret that we were compelled by time to leave the Campinha. In collecting, we had been more than usually successful. The hospitality of the Senhor had exceeded what we had seen, even in this hospitable country. His kindness followed us to the last moment, for we found that, without our knowledge, he had sent to the boat a store of roasted fowls, and other provisions, not the most lightly esteemed of which, were, some bottles of choice old Port, that had not seen the light for many a long year.

We left, intending to go below Vigia a few miles, and shoot Ibises, and for this purpose took one or two hunters with us in a montaria. As we passed the kingfisher bank, A—— took the montaria with Francisco, and upon overtaking us, an hour after, brought five of the larger and one of the small birds.

Six or seven miles below Vigia, we anchored at the entrance of a small igaripé, beyond which, the retiring tide had left exposed a broad sand beach. Here we anticipated finding plenty of Ibises, and forthwith, started A—— and the hunters, with as great expedition as though a flock of those birds were in full sight and waiting to be shot. I took the matter more leisurely, and sans ceremony, plunged into the surf, enjoying a luxurious bath, and finding plenty of amusement in netting Four-eyed Fish, that were in abundance along the edge of the water. Thereafter, I strolled along the beach for shells, but an hour's search gave me but one worth picking up. The water at this place is fresh during the rainy season, and salt in summer, and probably, shell-fish of either salt or fresh water do not flourish amid these changes. The blacks, meanwhile, were filling a basket with large crabs, which they found in deep holes in the mud, near shore. All the hunters returned unsuccessful, but reported Ibises, or Guerras, further down; and therefore, we prepared to go below in the canoe. During the day several Ibises had passed by, their scarlet livery, of dazzling beauty, glittering in the sunlight. As we coasted along in the dusk of evening, we could discover the beach, in many parts, black with sand-birds, that had collected for the night.

We were terribly annoyed, this night, by the sand-flies and small gnats, swarms of which seemed to have scented us out, and caused an intolerable itching. Morning found us anchored in an igaripé, and as soon as the tide would allow, we dropped below to the beach. The men again were unsuccessful, bringing in nothing but a young Spoonbill. It was now so late, and we had lost so much time, that we determined not to return to Vigia, where we had intended to pass a day or two; therefore, we bade adieu to our faithful hunters, feeling as much regret as if they had been friends of long acquaintance. A fair wind was blowing up the river, and the tide was favorable. The former soon became a tremendous gale, and the black clouds battled fearfully. The foresail was carried away, the blacks began to call on the Virgin, the frightened pilot forgot his helm, and nothing but the breadth of the canoe kept us from going under. A—— sprung to the helm, and in a moment, consternation gave place to effective alacrity, and we were safe. By ten o'clock, next morning, we were in Pará,

A letter from Senhor Gudinho to his wife, requested her to send us a singular pet animal, which the Senhor described as small, having a broad tail, with which, umbrella-tike, it shielded itself from the rain, and a lightning-like capacity for moving among the trees, now at the bottom, and, quicker than thought, at the top. But most curious of all, and most positively certain, this little quadruped was hatched from an egg. We suggested to the Senhor various animals, but our description of none answered. Of course, curiosity was at boiling point. We had heard of furred animals with ducks' bills, and hairy fish that chewed the cud; of other fishes that went on shore and climbed trees; of two-headed calves, and Siamese twins; but here, at last, was something unique—an animal hatched from an egg—more wonderful than Hydrargoses, and a speculation to make the fortunes of young men of enterprise. All day we waited, and nothing came; the next morning dawned, the noon bell tolled, and we, at last, concluded that the Senhora had been loath to part with so singular a pet, and that the instructions of her honored lord were to be unheeded. Dinner came, soup was on our plates, spoons were in our hands, and curiosity had expended itself by its own lashings, when a strange footstep was heard at the door-way, and a well-dressed, dusky Rachel appeared, bearing a carefully covered cuya intuitively to A———. Here was the wonder. What is it? What can it be? What is it like? Down went soup spoons; suspense was painful. First, unrolled a clean, little white sheet—second, another of the same,—the slightest possible end of a tail protruded from under a third, a little round nose and a whisker peeped from the remaining cotton,—and up leaped one of the prettiest little squirrels in the world. The little darling! Every body wanted him; every body played with him; and for a long time, he was the pet of the family, running about the house as he listed.

The Indians all believe that if they shoot at a squirrel, the gun is crooked ever after. Such superstitions are common with respect to other animals, and as they are harmless, deserve to be encouraged.

Chapter IX

FIRST DISCOVERY OF THE AMAZON BY PINZON—EXPEDITION OF GONZALO PIZARRO—DESCENT OF ORELLANA—SETTLEMENT OF PARÁ—SECOND DESCENT—ASCENT OF TEIXERA, AND ARRIVAL AT QUÍTO—HE DESCENDS WITH ACUÑA—INDIAN TRIBES—RIVERS, ETC.—THEIR REPORT OF THE COUNTRY—NUMBER OF TRIBES—INDIAN CUSTOMS—LANGUAGES—LINGOS GERAL—CANNIBALS—SYSTEM OF THE JESUITS—THEIR BANISHMENT—PRESENT SYSTEM, AND CONDITION OF THE INDIANS—THE GOVERNMENT—COMPULSORY LABOR

Before commencing the narrative of our Amazon expedition, a few particulars relating to the early history of this river may not be uninteresting. For these, I am in great part indebted to Southey, whose extensive work upon Brazil is the only one of authority readily accessible.

Seven years after the discovery of America, Vincente Yañez Pinzon, who, under Columbus, had commanded the Niña, obtained a commission from the Spanish sovereigns to go in search of new countries. The first point at which he arrived, is now called Cape St. Augustine, and here he landed, and took formal possession of the country. Coasting thence northward, the Spaniards came to what they called a sea of fresh water, and they supposed themselves in the mouth of some great river or rivers. It was the mouth of the Amazon. Without effecting further discovery, beyond landing at one of the islands, Pinzon continued on to the Orinoco, and thence returned to Spain. He believed that the land which he had visited was India beyond the Ganges, and that he had sailed beyond the great city of Cathay. This expedition carried many curious productions of the country, but none excited so much astonishment as an opossum, an animal unknown in the old world. It was described as having the fore part of a fox, the hind part of a monkey, the feet of an ape, and the ears of a bat; and was sent to Seville, and then to Grenada, that the King and Queen might see it. One or two other attempts were made to explore the vicinity of the entrance of the Amazon, within the next forty years, but without much success.

About the year 1541, Gonzalo Pizarro heard of a country rich in spices, to the eastward of Peru, and resolved to secure its possession. For this purpose, he set out from Quito, with about two hundred foot soldiers, one hundred horse, and four thousand Indians. Before they had advanced thirty leagues, they suffered extremely from earthquakes and storms, hunger and cold, At this distance. Pizarro was joined by the knight Francisco de Orellana, with a small reinforcement. Continuing on, the Spaniards suffered terrible hardships. The Indians died, or deserted, the soldiers wasted away, and, at last, upon the river Coca, they were in an excessive famine.

The Dorado of which they were in search, was as distant as ever, but still their hopes were fed by the delusive reports of the natives. To obtain relief, Pizarro sent forward Orellana, in a brigantine which they had built, with fifty men, and with orders to proceed to a fertile country, and to return as speedily as possible with provisions. Amid perils and disasters, the knight continued down about one hundred leagues, unto the river Napo. The country, through which he had passed, was uninhabited, nor was there any sign of culture or of population there. It was impossible to return, and if they waited for the army, they should perish with famine. Orellana conceived the adventurous hope of being himself the explorer of the great river, and his men were easily persuaded to acquiesce in his purpose. It was upon the last day of December, 1541, that the little band set forth. Sometimes, they met friendly Indians, at others, they were obliged to fight their way, sword in hand, through swarms of enemies. Famine and sickness thinned them. The river seemed interminable; still on, on. Hostile Indians increased in number; they were hardly ever out of sight of their villages. It was the 8th of August, 1542, when they sailed out of the river. They had built another brigantine upon their way, and now, the two were carried towards the West Indies by the current. Landing upon one of the islands, our adventurers proceeded thence to Spain. They had accomplished one of the most wonderful voyages ever made, and were received with distinguished honors. The account published by Orellana and the friar, who accompanied him, contained so many fabulous inventions, as to utterly destroy the authenticity of the whole. Not the least of these, was their account of a nation of Amazons which they had encountered, and which thereafter gave the river its name. Orellana received permission to repeat his discoveries, with a grant of dominion. Returning, he was unable to find the entrance of the river among the islands, and died worn out by vexation.

In 1615, Caldeira founded the city of Pará, and this was the first attempt by the Portuguese to colonize the river. The Dutch had previously formed a settlement upon the northern bank, some leagues above; but being soon driven out, the Portuguese remained sole masters.

In 1637, the Amazon was descended, a second time, by two ecclesiastics and six soldiers. They had formed part of a large deputation sent to Christianize the Indians, upon the frontiers of Peru, and meeting nothing but danger in their undertaking, had preferred the descent to the prospect of certain death in returning.

These fathers were so stupefied with fear, as to be unable to give any intelligible account of what they had seen, except horrible narrations of cannibal Indians. They were treated most courteously by the Governor of Pará, and in sending them home, that officer availed himself of the opportunity to cover his usurpation of the magistracy of the province, by an offer to do the State service in exploring the river. His proposition was approved, and Pedro Teixera was appointed commander of the expedition. He left Pará, the 28th of October, 1637, with seventy soldiers, and twelve hundred native bowmen and rowers, making with their women and slaves, two thousand persons in all, and embarked in forty-five canoes. The adventurers arrived, late in the succeeding year, at Quito, and their advent was celebrated by processions and bullfights.

The journal and map of Teixera were dispatched to the Viceroy of Peru, and this officer ordered Teixera to return, taking competent companions, who should survey the river, and prepare a report of its wonders for the Court at Madrid. Two professors were chosen for the purpose, Acuña and Artieda, and from their published narrative, we have the first authentic accounts of the Amazon. Embarking upon one of the small streams near Quito, the party soon arrived at the Napo. Here they encountered a tribe of Indians, called Encabellados, or long haired; so called from the custom with both sexes, of suffering their hair to reach below the knees. They were formidable enemies, and were constantly at war with neighboring tribes. They were cannibals; and in battle, their weapon was the dart. Further down, was the country of the

Omaguas, or flat-heads; whose peculiar custom resembled that of certain tribes of North American Indians. This was the most civilized, rational, and docile tribe upon the whole river. They grew and manufactured cotton, and made it an article of traffic with their neighbors. From this tribe was first learned the use of the seringa or rubber. They possessed the islands in the river for an extent of two hundred leagues, and were constantly warring with the Urinas on the south side, and the Tucunas on the north. The latter of these believed in metempsychosis, and worshipped a household idol. They were clothed about the loins with the bark of a tree; and were remarkable for their skill in stuffing birds, which they shot with the blow-gun. The Urinas were cannibals; shaved the crown of the head; and wore feathers of macaws in the corners of their mouths, besides strings of shells pendent from cars, nostrils, and under-lip.

Passing many other curious tribes, differing in customs and character, our adventurers came to the country of the great tribe called Curiciraris, who possessed an extent of eighty leagues, in the vicinity of the river now called Jumá. Their settlements were almost continuous. They were the shyest tribe upon the river, but among the most improved. They were excellent potters, making not only jars and pans, but even ovens and frying-pans, and in these they trafficked with other tribes. Here were first perceived golden ornaments, and Teixera was assured of a river of gold, running from the mountains, some days' journey to the northward.

Not far below, was the great river Jupurá, so called from a tribe of Indians thus denominated from a fruit of which they made a black paste for food. This river is one of the greatest tributaries of the Amazon.

The next considerable river was the Puros, named also from the tribe upon its banks. Here Teixera heard of a tribe of enormous giants, dwelling two months' voyage up the river. The Puros, were remarkable for their expiatory fasts, during which no state of infirmity or disease was admitted as a relaxation, and numbers actually died of abstinence from food.

Below the mouth of the Puros, upon the southern side, were the Caripunas and Zurinas, tribes remarkable for their skill in carving.

The next river, of note, was the Rio Negro. Here were rumors of remote people wearing hats and garments, and the voyagers concluded that this fashion was learned in consequence of their vicinity to some Spanish city. They also heard of a great river to the north, communicating, by a branch, with the Rio Negro. This was the Orinoco, but geographers were long incredulous as to the existence of such a connection.

The next great river was the Madeira; so named from the great quantities of wood floating down its current. Twenty-eight leagues below, was a great island, possessed by the Tupinambas, and called after their name. This tribe reported their ancestors to have emigrated from the region of Pernambuco, to escape the Portuguese. They were expert archers. They reported two remarkable races upon the southern shore; one of whom, were dwarfs, not bigger than little children; and the others, singular from their feet, which grew backwards. They also reported the existence of a nation of Amazons, and gave minute details of their appearance and habits. Whether such a nation ever existed or not, can never be ascertained; but it is most remarkable, that almost every tribe throughout Brazil, even those most separated, and speaking entirely different languages, should have believed in their existence. When Condamine descended the river, in 1743, he omitted no opportunity of inquiring after the Amazons, and invariably received the same reports.

Below the island of the Tupinambas, about eighty leagues, was the river Topajos, named from the tribe so denominated. These Indians were dreaded by the Portuguese; for their arrows were venomed with so powerful a poison, that the slightest puncture occasioned inevitable death. Here were Portuguese settlers, and a fort, on the present site of Santarem. Continuing on, our voyagers passed many lesser rivers; and heard rumors of gold and diamonds, far in the interior.

They arrived in Pará, upon the 12th of December, 1639; having scarcely met with an accident, end having enjoyed a most delightful voyage. They represented the country, through which they had passed, as rich beyond belief, capable or yielding all tropical productions; the forests, filled with wild animals and game; and the river, teeming with fish and turtle. Every where, were inestimable gums and drugs; and for ship-building there were timbers of the greatest strength and beauty.

The number of tribes, were estimated at one hundred and fifty, speaking different languages; and bordering so closely, that the sound of an axe in the villages of one, might be heard in the villages of another. Their arms, were bows and arrows; their shields, of the skin of the cow-fish, or of plaited cane. Their canoes, were of cedars, caught floating in the stream; their hatchets, were of turtle-shell; their mallets, the jaw-bone of the cow-fish; and with these, they made tables, seats, and other articles of beautiful workmanship. They had idols of their own making, each distinguished by some fit symbol; and they had priests, or conjurers. They were of a less dark complexion, than other Brazilian nations; were well made, and of good stature, of quick understanding, docile, disposed to receive any instruction from their guests, and to render them any assistance.

The Amazon, in its natural features, is the same now, as when Acufia descended; and the rapturous descriptions which he has given of these wild forests, and mighty streams, might have been written to-day. But where are the one hundred and fifty tribes, who then skirted its borders, and the villages so thickly populated?

Most of the Brazilian Indians, spoke languages somewhat resembling each other. The Tupi, in its dialects, prevailed in Brazil; as the Guanmi, in Paraguay; and the Omagua, in Peru. Of these three, the second is the parent, as the Greek is of the Latin. The Jesuits, in Brazil, adopted the Tupi; and this, under the present name of the Lingoa Geral, or general language, is understood by every Indian. Still, each tribe has its own peculiar dialect; and those, in contact with the whites, speak also the Portuguese.

The Tupi races were cannibals, and it was only after long and unwearied exertion, that the Jesuits could succeed in abolishing that practice. Rumor speaks still of cannibal Indians; but we never were able to obtain any account of such tribes, as deserved a moment's credence.

The Jesuits were always the firm friends of the Indians, and entertained the noble conception of civilizing and Christianizing those unnumbered millions of wild men, and of elevating them, within a very few generations, to a rank with other nations of the earth. They gathered them in villages, taught them the lingoa geral, and instructed them in arts and agriculture. They opposed, most determinedly, the enslaving of the Indians, and the cruelties of the whites. The Carmelites as resolutely defended the colonists, and the history of this province, for a long course of years, is little more than the detail of the struggle between these rival orders. The Monks were victorious; the Jesuits were forced to leave the country, and were transported like felons to the dungeons of Portugal. Their property in Brazil was confiscated, and at this moment, there is scarcely a public edifice in the province of Pará, but that belonged to them. The Government undertook to carry out the beneficent plan of the Jesuits, and, for this purpose, sent Friars throng the wilderness, to collect together the Indians, and offered them the rights of freemen. But partly owing to the inefficiency of the means, and partly to obstructions thrown in the way by the colonists, the system introduced by the Government proved ineffectual in preventing the diminution of the tribes, or in materially bettering the condition of the few who were willing to embrace its offers. Although nominally freemen, they are really the slaves of any white man who settles among them, and this must be the case, so long as they feel their real inferiority. The only hope for them is, that, in course of a few generations, their race will be so amalgamated with that of the whites as to remove all distinction. But, as far as our observations extended, their condition was superior, morally, to that of the frontier Indians in North America.

The head men, or chiefs, of the different settlements are denominated Tauçhas, and have the rank, and wear the uniform of, Colonels in the Brazilian service. In each district is also a Capitan des Trabalhadores, or Captain of the Laborers, and to him belongs the general supervision of the Indians and free negroes. If a certain number of men are required to navigate a vessel, or for any other purpose, the Capitan sends a requisition to the Tauçha, and the men must be forthcoming, no matter what may be their private engagements. This looks very like compulsion, but it is really no more so than jury duty. The men make a voyage to the city and back, and are then discharged, perhaps not to be recalled for several months. They are paid stipulated wages, and rations, and are sure of good treatment; for, besides that they have their own remedy, by running away, which they will do upon the least affront, the law throws over them strong protections. While we were at the Barra of the Rio Negro, a white man was lingering out a three years' imprisonment, for merely striking an Indian in his employ. The Government has been sometimes severely strictured for its conduct towards the Indians, but it is difficult to see what more it could do for them than it has done.

Chapter X

PREPARATION FOR ASCENDING THE AMAZON—OUR COMPANIONS—THE GALLIOTA—INDIANS—PROVISIONS—DIFFICULTIES AT STARTING—DETAINED AT SR. LIMA'S—INCIDENT—AN AFTERNOON UPON THE BEACH—ANOTHER SITIO—MARAJO—THE TOCANTINS—ISLANDS—CIGANAS AND OTHER BIRDS—WOOD SCENE—HABITS OF OUR INDIANS—ARRIVE AT BRAVES—POTTERY PAINTING—WATER-JARS—FILING THE TEETH—FUNERAL OF A CHILD—A PALM SWAMP—SERINGA TREES AND GUM COLLECTORS—SLOTH—HOWLING MONKEYS—AN ADVENTURE—ENTER THE AMAZON—A MACAW HUNT

It was no easy matter to put all things in readiness for an expedition up the river. It was like preparing for a family movement to the Oregon. In addition to Mr. Bradley, two other gentlemen were to accompany us: Mr. McCulloch, the proprietor of a saw-mill at the Barra de Rio Negro, who had lately come down, with a raft of cedar boards, to within a few days' sail of the city; and Mr. Williams, a young gentleman from Newark, New Jersey, staying like ourselves at Mr. Norris's, and who desired a further acquaintance with the wonders of the interior.

The boat in which we were to make our cruise was called a galliota, a sort of pleasure craft, but well adapted to such excursions. It was thirty feet in length, having a round, canoe bottom, and without a keel. Its greatest width was seven feet. The after part was a cabin, lined on either side, and at the remote end, with lockers, for provisions and other matters. Upon each locker was scanty room for one sleeper, and two could lie comfortably upon the floor, while another swung above them in a hammock. In front of the cabin door was a tiny deck, and beyond this, covering the hold, and extending to within two feet of the extreme bow, was the tolda, covered with canvass, and intended for the stowage of goods or baggage. On either side of this tolda was a space, a foot in width, and level. Here, in most awkward positions, were to sit the paddlers.

These were Indians, mostly of the Mura tribe, heretofore spoken of as the worst upon the river. They were from a little village below the Rio Negro, and consisted of a Tauçha and five of his sons, the eldest of whom, the heir apparent, had his wife and two small children in the bow. Beside these, was a pilot, and three others, making altogether eighteen persons.

The after part of the cabin, and the whole tolda, with barely room enough for our trunks, and the fish and farinha for the crew, were crammed with Bradley's goods, bringing the deck within a few inches of the water.

Our main stock of provision was to be laid in at Pará, and the lockers, and every spare corner was occupied in their stowage. We had a couple of hams, great store of ground coffee, tea, sugar, coarse salt, onions, sardines, oil, vinegar, molasses, candies, tin cases of cheese, and two large bags of oven-dried bread. Sundry demijohns of wine and cashaça comprised the stock of drinkables, the former being for home consumption, the latter for rations to the crew. In addition to these things, several of our lady friends had contributed huge loaves of cake, and Yankee dough-nuts, and jars of doces, not a few. Not the least acceptable, were some pots of New York oysters, from a clever captain in the harbor.

We did not anticipate that a forty days' passage in this overloaded boat would be without all sorts of inconveniences; but such an adventure had charms enough, and we were determined to have a jolly cruise, the household gods *nolentes volentes*, as General Taylor would say.

No vessel can pass the fort at Gurupá without a permit from the authorities at Pará, and all voyagers on the river must provide themselves with passports. These we obtained without difficulty, and at slight expense. Doctor Costa, Mr. Campbell, and other friends, furnished us with letters to persons of note in the different towns which we were to pass.

At last, upon the 23^d of May, we were fairly on board, and ready to start with the tide. But here occurred a difficulty, and an ominous one, at the outset. Six of the Indians had given us the slip, not caring to return thus soon to the Rio Negro. Our remedies were patience and police, and we resigned ourselves to the one, hunting the runaways with the other. Towards night, they were brought in, and now, going on board again, we moored outside of a large canoe, to prevent a like disaster, and waited the midnight tide. Rain poured furiously, but we gathered ourselves around a trunk-table, and ate and drank long life to our friends, and a pleasant passage to ourselves. The Indians huddled about the door, feasting their eyes, and muttering their criticisms, but their envy was speedily dissipated by a distribution of cashaça, and biscuit, with a plate of oysters to the Tauçha. The old fellow bore his honor king-like, and I fancy, was the first South American potentate that ever tasted Dawning's best.

There was still opportunity for a short nap before the tide would serve, and we awaked just in time; but now was another trouble. The Indians, having no fear of wholesome discipline before their eyes, were desperately determined not to be awaked, and but for the ruse of calling them to a "nip" of cashaça, we might have lost the tide again. The effect was electrical, and they started from their deep slumbers, each striving to be foremost. There was one boy, however, who skulked into a montaria behind the large canoe, and would only be induced to come on board again by the capture of his trunk. Five on a side, they took their places. The Tauçha planted himself on the top, having a proper idea of prerogative, the children hid themselves away among the farinha baskets, and the princess covered herself in the bow, and prepared to sleep.

Our course was the same that we had formerly taken towards Caripé, and, by noon, we had arrived at the house of Sr. Lima, a trader, within two miles of that place. Here we stopped, not caring to pass the bay of Marajo by night, and improved the opportunity to make a sail. As the tide rose, towards night, word was brought that the galliota was leaking at such a rate to endanger the goods. No alternative was left but to unload her with all speed, and it was only by the most active exertions that she was kept from swamping. All the goods were piled in the verandah, and the lady of the house allowed us the small chapel, in which to dry some of the articles. We sent her a box of sardines, in token of our gratitude, and it seemed to unlock her heart chambers, for forthwith appeared a servant to attend our table, bringing a silver teapot, and various other appliances for our comfort. Slinging our hammocks in the verandah, about the goods, we slept in the open air. During the night, we were startled by a singular incident, trivial enough in itself, but one that carried us back to home scenes. Some voyager passed us, singing an air frequently sung in Sunday schools, at home, and known as the "Parting Hymn." We little thought, when last we heard it hymned by a congregation of children, that we were next to listen to it upon the far distant waters of the Amazon. The words were not distinguishable. We started the same tune in return, but the voyager was already beyond the reach of our voices, and lost behind a point of the island. Who this could have been, we were unable to ascertain at Pará, upon our return. It was not an American.

Repairing the galliota detained us two days, but every thing being carefully repacked, and the boat cleansed, we were amply repaid. Starting again, on the 25th, we hoisted our new lug-sail, and a fine breeze soon swept us past Caripé, our old shelling ground. Full tide forced us lo lie by at noon, and we brought up under a high bank, upon which was a sideless hut, containing a woman and children. The rest of the family, it being Sunday, had gone off to a fiesta in the neighborhood. The first impulse of the Indians, upon reaching shore, was to look out for some shade where they might stretch themselves to sleep. One or two of the more active, however, started out with a gun, and, before long, returned with a live sloth, which they had obtained by climbing the tree upon which he was suspended. This was of a different species from those we had seen near Pará. The beach was broad and sandy, and we amused ourselves with bathing, and searching for flowers, and seeds thrown up by the tide. Among the flowers was one most conspicuous, of the Bignonia family, large, yellow, and sprinkling in profusion the dark green of the tree which it had climbed. Wandering on some distance, we found ourselves in a little cove, secluded from the sunlight by a high, rocky bank, and so dark that bats were clustering about the tree trunks in numbers. The temptation was too strong, and we imitated the good example of the Indians.

By sunset, we were again pressing on, and, in the early evening, coasted along several miles. The shore hereabouts was lined with ragged sand rocks, and in case of squalls, which occur almost daily, during the rainy season, the navigation is hazardous. Our own situation began to cause us some anxiety. Several times the bottom of the galliota had scraped upon the rocks, and we were only forced off by the Indians springing into the water, and dragging us free. A storm was gathering, and vivid lightning and low growling thunder betokened its near approach. A man, at the bow, constantly reported the water more and more shallow, and the rising waves dashed hoarsely upon the near rocks. But just then a little igaripé opened its friendly arms, and, almost in a moment, we were beyond harm's reach, in water calm as a lake.

The morning dawned pleasantly, and a fine breeze springing up, we soon crossed the bay, and, by noon, had arrived at a nice beach, upon which was a grove of assai palms loaded with fruit. Here we stopped to fill our panellas. Continuing on a few miles, we struck into a narrow channel, and came to an inviting-looking house, where we concluded to wait the gathering storm. The occupants were two Brazilians, of a better class than we had seen since leaving the city, and we were received with warmth. The frame of the house was covered entirely, even to the room-partitions, by the narrow leaves of a species of palm, plaited with the regularity of basket work. A quantity of cacao lay drying upon elevated platforms, and around the house hung much dried venison. Deer were abundant here, and one had been killed that morning. But what gratified us most was a goodly flock of hens, and we at once commenced a parley for a pair, for we had become somewhat tired of ham. Meanwhile the women had been preparing our assai.

The region of country that we were now in, was exceedingly low, mostly overflowed at high water. The waters had fallen about a foot, but still, every thing around this house was wet, and we had only gained access to it by walking from the boat on logs.

The next day, the 27th, we coasted along Marajo, observing many novel plants and birds. One species of palm, particularly attracted attention; its long feather-like leaves growing directly out of the ground, and arranged in the form of a shuttlecock. There now began to be great numbers of macaws, red and blue, flying always in pairs, and keeping up a hoarse, disagreeable screaming. We passed what was formerly a large and valuable estate, still having fine-looking buildings and a chapel. It had belonged to Mr. Campbell, and like many an other, had been ruined during the revolution of '35.

We crossed the mouth of the Tocantins, but without being able to discern either shore of that river. It appeared a broad sea, every where dotted with islands. The Tocantins is one of the largest Amazon branches, and pours a vast volume of water into Marajo Bay. This particular portion of that Bay is called the Bay of Limoeiro, and is crossed by vessels bound to Pará from the Amazon, in preference to the route which we had taken. The Tocantins, and a few small streams nearer the city, are often considered the legitimate formers of Pará river. But, through numerous channels, a wide body of water from the Amazon sweeps round Marajo, and the Gram Pará is a fair claimant to all the honors of the King of Waters.

The Tocantins is bordered by many towns, and is the channel of a large trade. The upper country is a mineral region, and diversified by beautiful mountain scenery. The banks yield fustic, and numerous other woods valuable as dyes, or for cabinet work, and if the efforts to establish a sawmill, now in contemplation, be successful, these beautiful woods will soon be known as they deserve. Great quantities of castanha nuts also come down the river. The town of Cametá, between thirty and forty miles from its mouth, contains about twenty-five hundred inhabitants, and is in the midst of an extensive cacao-growing region. This was the only town upon the Amazon that successfully resisted the rebels in 1835. The Tocantins is navigable for steamboats or large vessels for a great distance.

Since the 26th, we had been sailing among islands, often very near together, and again, several miles apart. Upon the 28th, we were unable to effect a landing until noon, so densely was the shore lined with low shrubs. Upon these sat hundreds of a large reddish bird, known by the name of Cigana, and common upon the whole Amazon, the Opisthocomus cristatus (Lath).

Among them, were numbers of bitterns, and a large, black bird, the Crotophaga major. This bird is often seen in flocks among the bushes which skirt the river, and is conspicuous for its long, fan-like tail, and graceful movements. Sometimes it is seen domesticated. There is another species, the C. ani, seen about the cattle, on the plantations; of smaller size, and inferior beauty. We afterwards obtained the eggs of the former, among the bamboos, at Jungcal. They were large, almost spherical, of a deep blue color, but covered entirely with a calcareous deposit, as are the eggs of many of the cormorants.

Having reached a spot where the bank was a little higher than elsewhere, we landed. A small opening between the trees allowed ingress, and we found ourselves in a fairy bower. How much we longed for the ability of sketching these places, so common here, so rare elsewhere. Not the least interesting feature was the group of Indians about the blazing fire, some attending to their fish, which was roasting on sticks, inclined over the flame; others sitting listlessly by, or catching a hasty nap upon their palm leaves. A tree bearing superb crimson flowers shaded the boat, and a large blue butterfly was continually flitting in and out among the trees, as if sporting with our vain attempts to entrap him. Not far off, macaws were screaming, and the shrill whistle, observed in the woods near Pará, sounded from every direction.

We had now been nearly a week in the galliota, and although somewhat crowded, had got along very comfortably. The only inconvenience was the sultry heat of the afternoon; for, in these narrow channels, the wind had little scope. But no matter how severe the heat, the Indians seemed not to mind it, although their heads were uncovered, and their bodies naked. Every day, about noon, they would pull up to the bank for the purpose of bathing, of which they were extravagantly fond. Even the little boys would swim about like ducks. Their mother, the princess, had quite won our esteem, by her quiet, modest demeanor. Her principal care was to look after the children, but she spent her spare hours in making cuyas from gourds, or in sewing for herself or her husband. He, good man, seemed very fond of her, which would not have been surprising, except in an Indian; and always paddled at her side. He might have been proud of her, even had his potentacy expectant been more elevated, for she was very pretty, and her hands and arms might have excited the envy of many a whiter belle.

Early upon the 29th, we arrived at Braves, a little settlement, where was lying Mr. McCulloch's raft. Upon this was stationed a "down east" lumberman, by name Sawtelle, who was to add another to our full cabin. We were to remain at Braves until the arrival of a large vessel, or battalon, which was engaged in the transportation of the boards; and as this was likely to be some days, we unloaded upon the raft, slung our hammocks under the thatched cabin, and sent the galliota, again badly leaking, to be recaulked.

Braves is one of the little towns that have grown up since the active demand for rubber, of which the surrounding district yields vast quantities. It is a small collection of houses, partly thatched, and partly of mud, stationed any where, regardless of streets or right lines. Bradley and I started to explore for eggs whereon to breakfast. We found our way to a little affair called a store, or venda, in front of which, a number of leisurely gentlemen were rolling balls at a one-pin. We were politely greeted with the raised hat, and the customary "viva," and a chance at the pin was as politely offered, which with many thanks, we were obliged to decline. Our errand was not very successful, for upon the next Sunday was to be a fiesta in the vicinity, and the hens were all engaged for that occasion. At one of the houses, an old Indian woman was painting pottery, that is, plates, and what she called "pombos" and "gallos," or doves and cocks, but bearing a very slight resemblance to those birds. Another was painting bilhas, or small water jars, of white clay, and beautiful workmanship. She promised to glaze any thing I would paint, giving me the use of her colors. So I chose a pair of the prettiest bilhas, and after a consultation on the raft, we concluded to commemorate our travels by a sketch of the galliota. It was a novel business, but after several trials I made a very fair picture, with the aid of contemporary criticisms. The old Tauçha was mightily pleased to see himself so honored, as were the others, who gathered round, watching every movement of the pencil, and expressing their astonishment. The figure of the princess especially excited uproarious applause. Beside these, were several other devices, and at last, all complete, I took my adventure to the old woman. But she was provoked at something, and would not be persuaded to apply the glazing. However, after much coaxing and many promises, she assured us that we should have them on our return down the river. The colors she used were all simple. The blue was indigo; black, the juice of the mandioca; green, the juice of some other plant; and red, and yellow, were of clay. The brushes were small spines of palms, and the coloring was applied in squares or circles; or, if any thing imitative was intended, in the rudest outline. The ware was glazed by a resinous gum found in the forest. This was rubbed gently over, the vessel previously having been warmed over a bed of coals.

The stream opposite Braves, was one-fourth of a mile wide; and beyond, was an island heavily wooded. Thither we sent a hunter every day, and he usually brought in some kind of game; a Howling Monkey or macaw. For ourselves, we were, confined pretty much to the raft; the region about the town being nothing but swamp. Yet still, we found opportunity to increase our collection of birds by a few specimens hitherto unknown to us, particularly the Cayenne Manikin, and the Picus Cayanensis. The Indians, meanwhile, had found a quantity of rattan, and were busily engaged in weaving a sort of covering, or protection from the rain. Two long, cradle-shaped baskets were made, one fitting within the other, the broad banana leaves being laid between; and under this, they could sleep securely.

We were struck, at Braves, by the appearance of some Portuguese boys, whose teeth had been sharpened in the Indian manner. The custom is quite fashionable among that class, who come over seeking their fortunes; they evidently considering it as a sort of naturalization. The blade of a knife, or razor, is laid across the edge of the tooth, and by a slight blow and dexterous turn, a piece is chipped off on either side. All the front teeth, above and below, are thus served; and they give a person a very odd, and to a stranger, a very disagreeable appearance. For some days, after the operation is performed, the patient is unable to eat or drink without severe pain; but soon, the teeth lose their sensitiveness, and then seem to decay no faster than the others.

One day, there was a funeral of a child. For some time previous to the burial, the little thing was laid out upon a table, prettily dressed, and crowned with flowers. The mother sat cheerfully by its side, and received the congratulations of her friends, that her little one was now an angel.

On the morning of June 1st, we were delighted to see the battalon come swiftly up with the tide, and made immediate preparations for departure. Now, was trouble again with the Indians. Some of the Tauçha's boys wanted to return to Pará, and the old fellow evidently did not care whether they did or no, notwithstanding his oft-repeated assurances, that he would keep them in order. His authority was very questionable, and we were getting tired of his lazy inefficiency. The old remedy was tried, and again we were conquerors. These difficulties are incident to every navigator upon the river; for, upon the slightest whim, an Indian is ready to desert, and often, the detention of their little baggage, or the wages accruing to them, is matter of perfect indifference.

The morning of the 2^d, found us in a narrow stream, winding among small islands, which were densely covered with palms. Landing, in what was almost entirely a palm swamp, we amused ourselves a long time, by observing the different varieties, of which we had no means of ascertaining the name, and in collecting the fruits. Here were numbers of the shuttlecock palms; and their large leaves, spread upon the wet ground, made the Indians a comfortable bed. There are more than one hundred described species of palms, in Brazil; growing, to some extent, almost every where. But, within the province of Pará, by far the larger portion are upon the islands, at the mouth of the river. Upon the islands above, and upon the main-land, they are comparatively rare.

Leaving the palms, we came to a region abounding in huge trees, where the shore was every where easy of access. Here were numbers of seringa trees, and we passed many habitations of the gum collectors. These were merely roofed, or thatched on one side, and very often the water rose to the very door. No fruit trees of any sort were there, nor was there sign of cultivation. The forest around was just sufficiently cleared to avoid danger from falling trees, or to let in a glimpse of the sun. In these miserable places were always families, and thus they live all the year round, eating nothing but fish and farinha, and their situation only bettered in summer by less dampness.

We now entered one of the direct channels from the Amazon, called the Tapajani. It was half a mile in width, and through it poured a furious current. Here we saw a Sloth, climbing, hand over hand, up an assai palm, by the water; and here, also, we first heard in perfection the Guariba, or Howling Monkey. There were a number of them, some, near by, and others, at a great distance; all contributing to an infernal noise, not comparable to any thing, unless a commingling of the roaring of mad bulls, and the squealing of mad pigs. This roaring power is owing to the peculiar conformation of the bones of the mouth, by which they are distinguished from all others of the family. We got quite up to a pair of these fellows, as they were making all ring, deafening even themselves. They were in a tree-top close by the water, and a shot from A—— brought down one of them. But recovering himself, he made off, as fast as he was able through the bushes. Immediately the boat was stopped, and A——, with several of the Indians, sprang on shore in pursuit, but without success. There were still some young ones in the tree, and another shot sent tumbling one of these. But he too saved himself, twisting his tail about a limb as he fell, and, in a twinkling, he was snug in a corner, safe from our eyes. Monkey hunts often end so.

Leaving the Tapajani, we were still separated from the main current of the Amazon, by a long island, two or three miles distant, and it was noon of the 5th, before, through the space intervening between this and an island above, we were able to distinguish the northern shore, twenty miles away. The bank near us was bold, and evidently the force of the current was continually wearing upon it and undermining the enormous trees, that towered with a grandeur befitting the dwellers by this unequaled river. Often, the boat struck upon some concealed limb or trunk, usually only requiring us to back off, but sometimes, making us stick fast. In such cases, several of the boys would jump into the water, and in a great frolic, drag us free.

Towards evening, we came to a place where the macaws were assembling to roost. Disturbed by our approach, they circled over our heads in great numbers, screaming outrageously. A——

caught a gun, and as one of them came plump into the water, winged, Tauçha, men, women and children set up a shout of admiration. Two of the boys were instantly in the stream, in chase of the bird, who was making rapid strokes towards a clump of bushes. Macaw arrived first, and for joy at his deliverance, laughed in exultation; but a blow of a pole knocked him into the water again, and a towel over his nose soon made him prisoner upon our own terms. The poor fellow struggled lustily, roaring, and using bill and toes to good purpose. His sympathizing brethren flew round and round, screaming in concert, and it was not until another shot had cut off the tail of one of the most noisy, that they began to credit us for being in earnest. Our specimen was of the Blue and Yellow variety. During the night, we repeatedly sailed by tree, where these birds were roosting, and upon one dry branch, A——, whose watch it was, counted eighteen. The opportunity was tempting, but we were under press for Gurupá, and could not delay. The Indians were as anxious for a rest as ourselves, and all night, pulled, with scarcely an intermission.

Chapter XI

***ARRIVE AT GURUPÁ—SITUATION OF THE TOWN—
RECEPTION BY THE COMMANDANT—AN EGG HUNT—
STORM—CROSS THE XINGU—CÁRAPANÁS—CEDAR
LOGS—HARPY EAGLE—BIRDS—MOUNTAINS—INDIAN
COOKING—FOREST TREES—SNAKE BIRDS—A
TOUCAN'S NEST—MUTVCAS—INDIAN IMPROVIDENCE—
GRASS FIELDS—ENTER AN IGARIPÉ—HYACINTHINE
MACAWS—PASSION FLOWERS—PASS PRYINHA—
MONTE ALÉGRE—ARRIVE AT SITIOS—THRUSH—
CAMPO—INCIDENT—ENTER THE TAPAJOS—WHITE
HERONS—FLOWERING TREES—ARRIVAL AT
SANTAREM—CAPT. HISLOP—MORNING CALLS—
BEEF—RIVER TAPAJOS—FEATHER DRESSES—
EMBALMED HEADS—DESCRIPTION OF SANTAREM—
DEPARTURE—A SLIGHT DIFFICULTY***

Early on the 6th, Gurupá was in sight. As we drew near, we were hailed from the fort in some outlandish tongue, inquiring, probably, if we intended to storm the town. Our answer was in English, and they seemed as wall satisfied as though they had comprehended it, bidding us pass on. The town does not present a very striking appearance from the water, merely the tops of half a dozen houses being visible. The landing was at the upper end, and there we moored, among numbers of little craft which had collected from the vicinity, for the day was a fiesta.

Gurupá was formerly considered the key to the river, and was of great service to the early colonists in preventing the encroachments of other nations. Now, it is of little consequence, and has but a scanty trade. Its population numbers a few hundred. Superior sarsaparilla, or salsa, is taken to Pará from this vicinity. The situation of the town is fine. In front, a long island stretches far down the river, called the Isle of Paroquets. Above, and within a few miles, are two other islands, both small, and beautiful from their circular shape. Upon the Isle of Paroquets, all kinds of parrots and macaws were now preparing to breed, in vast flocks, and this accounted for the unusual numbers which we had seen, within a few days.

We had a letter from Doctor Costa to the Commandant, and suitable respect, moreover, demanded a display of passports; so after breakfasting on the beach, A—— and Bradley went up to his Excellency's house. The Commandant was very polite in his attentions, and pressed us strongly to remain to a dance, which he was to give in the evening. but if we could only wait until afternoon, he would send us some fresh beef; and, at any rate, upon our return, we must stay with him at least a fortnight. While our two diplomatists were thus engaged, Sawtelle undertook the customary search for eggs; and the first person he made inquiry of for these indispensables, was the schoolmaster, who with his dignity all upon him, and his scholars about him, was discharging his usual duties. Yes, the schoolmaster had eggs, and at once started to bring them, careless of dignity, duties and all. In his absence, our messenger dispatched the scholars to their respective homes, on a like errand, and soon, they returned with one, two, and three apiece, until our cuya was filled. There are no County Superintendents, or Boards of Trustees, in Brazil.

A fresh breeze had sprung up, and we hasten away. A few miles above Gurupá, the clouds began to darken, the waves were rising ominously, and there was every appearance of a squall: several canoes, which had been on the same course, had hauled in shore, and their crews seemed to look upon us with astonishment, as we swept by them. A—— was on deck as usual, watching the sail, and the Indians, half frightened at our speed, kept every eye on him. Suddenly a halyard parted, the sail flaunted out, the boat tipped, and there was not an Indian on board but crossed himself, and called on Nossa Senhora. Perhaps Nossa Senhora heard them, and was willing to do them a good turn, for very soon the wind died away, and the bright sun made all smile again.

Soon after dark, we crossed the mouth of the Xingu, (Shingu), much to the displeasure of the Indians, who wished to stop upon the lower side. And they were very right; for scarcely had we crossed, when we were beset by such swarms of cárapanás, or musquitoes, as put all sleep at defiance. Nets were of no avail, even would the oppressive heat allow them, for those who could not creep through the meshes, would in some other way find entrance, in spite of every precaution. Thick breeches they laughed at, and the cabin seemed the interior of a beehive. This would not do, so we tried the deck; but fresh swarms continually poured over us, and all night long, we were foaming with vexation and rage. The Indians fared little better, and preferred paddling on, to anchoring near shore. The English consul at Pará had told us, "Ye'll be ate up alive intirely," and certainly this began to look much like it. Moreover, we were told for consolation, that this was but the advanced guard. It is very remarkable that cárapanás are not found to any troublesome extent below the Xingu. The country is low, and much of it wet, yet, from some cause, does not favor these little pests.

The Xingu is a noble river, in length nearly equal to the Tocantins. At its mouth, it expands to a width of several miles, and is there profusely dotted with islands. From the Xingu, the best rubber is brought, and a number of small settlements, along the banks, are supported by that trade.

Soon after sunrise, upon the 7th, we brought up along side of a large cedar log, the land being inaccessible, or rather, being entirely overflowed, and speedily, we had a rousing fire kindled between two of the roots. This cedar is a beautiful wood, light as pine, and, when polished, of fine color. Most of the woods of the country are protected against the ravages of insects, by their hardness, but the cedar is filled with a fragrant resinous gum, which every insect detests. It grows mostly upon the Japurá, and other upper branches of the Amazon, and is almost the only wood seen floating in the river. At certain points, along the shores, vast numbers of the logs are collected, and were mill streams common, might be turned to profitable purpose.

Just before we had reached our mooring, a full-sized Harpy Eagle perched upon a tree near the water, his crest erect, and his appearance noble beyond description. We gave him a charge of our largest shot, but he seemed not to notice it. Before we could fire again, he slowly gathered himself up, and flew majestically off. This bird is called the Gavion Real, or Royal Eagle, and is not uncommon throughout the interior. Its favorite food is said to be sloths, and other large sized animals.

After breakfast we sailed by a broad marsh, upon which hundreds of herons were stalking through the tall grass. Upon logs, and stumps projecting from the water, sat great flocks of terns. ducks and cormorants, who, at our approach, left their resting places, some, circling about us with loud cries, others, diving beneath the water, or flying hurriedly to some safer spot.

We proceeded very slowly. The current had a rapidity of about three miles an hour, and it was only by keeping close in shore, that we could make headway. The water of the Amazon is yellowish, and deposits a slight sediment. It is extremely pleasant to the taste, and causes none of that sickness, upon first acquaintance, that river waters often do. For bathing, it is luxurious.

Upon the morning of the 8th, a range of hills, or mountains, as they may properly enough be called, was visible upon the northern shore; and after passing such an extent of low country, the sight was refreshing. They had none of the ruggedness of mountains elsewhere, but rose gently above the surrounding level, like some first attempt of nature at mountain making.

We saw a number of Darters upon the branches over the water, but were unable to shoot them. A pair of red macaws fared differently, and we laid them by for breakfast. During the morning we passed about a dozen sloths. They were favorite food of the Indians, and their eyes were always quick to discover them among the branches, upon the lower side of which they usually hung, looking like so many wasp's nests. We observed a large lily, of deep crimson color, and numerous richly flowered creepers, but without being able to obtain them. It was impossible to effect a landing, and we moored again by the side of a cedar log, eight feet in diameter. Upon this was growing a cactus, which we preserved. Our macaws, fricasseed with rice, made a very respectable meal; somewhat tough; but what then, many a more reputable fowl has that disadvantage. The Indians shot a small monkey, and before life was out of him, threw him upon the fire. Scarcely warmed through, he was torn in pieces, and devoured with a sort of cannibal greediness, that made one shudder.

Palm trees had entirely disappeared, but cotton trees, of prodigious height and spreading tops, were seen every where. So also were mangabeira trees, conspicuous from their leafless limbs, and the large red seed pods which ornamented them. There was another tree, more beautiful than either, called from its yellowish brown bark, the mulatto tree. It was tall and slim, its leaves of a dark green, and its elegantly spreading top was covered with clusters of small white flowers. The yellow limbs, as they threaded among the leaves and flowers, produced a doubly pleasing effect. This tree is common upon the river, but its wood is esteemed of no value.

We made little advance, the wind not favoring, and the current being strong. Late in the evening, we threw a rope over a stump, at some distance from the shore, beyond reach of cárapanás, and spread ourselves upon the cabin top, in the clear moonlight, hoping for a quiet sleep. But the breeze freshened, and off we started again, to our great misfortune; for, the wind soon dying away, we got entangled in the cross currents, and were hurled with violence among bushes arid trees. And now a pelting storm came up, and the gaping seams of the cabin top admitted floods of water. To crown the whole, we were at last obliged to stop in shore, and sunrise found us half devoured.

We were always out as early as possible in the morning, for besides that it was far the pleasantest part of the day, there were always birds enough by the water side to attract one fond of a gun. The morning of the 9th was ushered in by a brace of discharges at a flock of parrots, and immediately after, down dropped a Darter. We had seen several of these within a few days, and they were always conspicuous from their long, snake-like necks, and out-spread tails. They were very tame, and easily shot; but. if not instantly killed, would dive below the surface of the water, with nothing but the tip of their bill protruding. In this manner, they would swim under the grass, and were beyond detection. The Indians called them Cararás. This family is remarkable for the absence of any tongue, save the slightest rudiment, and for having no external nostril. This specimen was a young male of the Plotus Anhinga.

We here saw another Harpy Eagle, and a variety of hawks; and in a large tree, directly over the river, was the nest of the Toco Toucan.

The land was still swampy, but we contrived to find a stopping place, where we were terribly persecuted by cárapanás. The hills, on our right, were increasing in number and size. Several canoes passed on their way down, but as these always keep in the current, one may sail the whole length of the Amazon, without hailing a fellow voyager. We were here annoyed by a large black fly, called mutúca, who seemed determined to suck from us what little blood the cárapanás had left.

The men rowed with a slight increase of unction, attributable to our being out of fish, which they had wasted in the most reckless manner. It was impossible to serve them with daily rations; no independent Indian would submit to that. No matter how large the piece they cut off, if it was more than enough for their present want, over it went into the stream. Of farinha too, they were most enormous gluttons, ready to eat, at any time, a quart, which swelling in water, becomes of three times that bulk. And they not only ate it, but drank it, mixing it with water, and constantly stirring it as they swallowed. This drink they called shibé.

The morning of the 10th, discovered the northern hills much broken into peaks, resembling a bed of craters. Many of the hills, however, were extremely regular, often shaped like the frustum of a cone, and apparently crowned with table land.

We coasted, for some hours, along a shoal bank, covered with willows, and other shrubs standing in the water. Such banks are generally lined with a species of coarse grass, which often extends into fields of great size. Large masses of this are constantly breaking off by wind and current, and float down with the appearance of tiny islets. A nice little cove invited us to breakfast, and the open forest allowed a delightful ramble. Soon after leaving this place, the channel was divided by a large island, and taking the narrower passage, all day we sailed southward, in what seemed rather an igaripé than a part of the Amazon. Here were thousands of small green, white-breasted swallows; and the bushes were alive with the Crotophagas, spoken of before. Here also we saw a pair of Hyacinthine Macaws, entirely blue, the rarest variety upon the river; and numbers of a new Passion flower, of a deep scarlet color. "In the lanceolate leaves of the Passion flower, our catholic ancestors saw the spear that pierced our Savior's side; in the tendrils, the whip; the five wounds in the five stamens; and the three nails, in the three clavate styles. There were but ten divisions of the floral covering, and so they limited the number of the apostles; excluding Judas, the betrayer, and Peter, the denier."

Re-entering the main stream, early upon the 11th, we passed the little town of Pryinha, upon the northern shore. The bank was still skirted by willows and grass, and the only landing we could discover, was in a swamp of tall callas. Upon the stems of these plants was a species of shell, the Bulimus picturata (Fer). There was here a large tree bearing pink flowers, of the size and appearance of hollyhocks; and crimson Passion vines were twined about the callas. During the day, we passed a number of trees formed by clusters of many separate trunks, which all united in one, just below the branches.

Upon the 12th, we passed Monte Alégre, a little town, likewise upon the northern shore, and noted above other river towns for its manufacture of cuyas some of which are of exquisite form and coloring. Just below the town, a fine peak rises, conspicuous for many miles. The shore, near us, was densely overhung with vines of the convolvolus major, or morning-glory, plentifully sprinkled with flowers of pink and blue. We passed a brood of little ducks apparently just from the shell. As we came near, the old one uttered a note of warning and scuttled away; and the little tails of her brood twinkled under the water.

About noon, discovering a sitio, we turned in, hoping to obtain some fish for our men, who grumbled mightily at their farinha diet. There were a couple of girls and some children in the house; and they seemed somewhat surprised at our errand, for they had not enough to eat far themselves. The poor girls did look miserably, but poverty in such a country was absurd.

Proceeding on, an hour brought us to another sitio, where the confused noises of dogs, and pigs, and hens, seemed indicative of better quarters. Here were three women only, engaged in painting cuyas. At first, they declined parting with any thing in the absence of their men; but a distribution of cashaça and cigars effected a wonderful change, and at last, they sold us a pig for one milree, or fifty cents, and a hen for two patacs, or thirty-two cents. Soon after, an old man from a neighboring sitio brought in a Musk Duck for one patac. We gave the pig to the men, and, in a few moments he was over their fire. Meanwhile, they caught a fish, weighing some dozen pounds, and with customary improvidence, put him also into the kettle. Finally, the half eaten fragments of both were tossed into the river. The old man, of whom we had bought the duck, was very strenuous for cashaça, and brought us a peck of coffee in exchange for a pint. Not content with that, he, at last, pursued us more than a mile, in a montaria, bringing eight coppers for more, and seemed to take it much to heart that we had none to sell.

Upon the 13th, we left the southern shore, in order to avoid a deep curve, and crossed to a large island. Coasting along this, we discovered a number of birds new to us, the most interesting of which was a small species of the Thrush family, the Donacobuis vociferans (Swain). This bird we often, afterwards, saw in the grass by the water, and his delightful notes reminded us of his cousin, the Mocking-bird, at home. He was incomparably the finest singer that we heard upon the river, and there, where singing birds are unusual, may be considered as one of the river attractions. Upon either side his neck, was a yellow wattle, by the swelling of which he produced his rich tones.

There was high land upon the southern shore, but upon our island we could find no place to rest. The Amazon, in this part of its course, expands to a width of from fifteen to twenty miles.

Towards night we bought a supply of dried peixe boi, at a sitio. It was inconceivably worse than the periecu, or common fish, in rankness and toughness.

We passed a campo extending back for several leagues, and covered with the coarse grass mentioned before, and mostly overflowed. This was said to be a place of resort for ducks, who breed there in the months of August and September, in inconceivable numbers. There were evidently many now feeding upon the grass-seed, and occasionally, a few would start up at the noise of our approach. Our pilot suggested that there were plenty of cattle and sheep upon this campo, and that they belonged to no one. The Indians were longing for fresh meat and had they been alone would have carried off one of the "cow cattle," as Bradley termed them, without inquiring for ownership.

During the morning of the 14th, we stopped at a cacao sitio, where was a fine house, and a number of blacks. While here, a montaria arrived, containing a sour-looking old fellow, and a young girl seated between two slaves. She had eloped from some town above with her lover, and her father had overtaken her at Monte Alégre, and was now conveying her home. She was very beautiful, and her expression was so touchingly disconsolate, that we were half tempted to consider ourselves six centuries in the past, toss the old gentleman into the river, and cry St. Denis to the rescue. Poor girl, she had reason enough for sadness, as she thought of her unpleasant widowhood, and of the merciless cow-hide in waiting for her at home. Some one asked her if she would like to go with us. Her eyes glistened an instant, but the thought of her father so near, soon dimmed them with tears.

All day we continued along the islands. Upon the southern shore, a range of regular highlands extended up and down, and along them, we could distinguish houses, and groves of cacao trees.

Towards evening, we passed a campo of small extent, having a forest background, and lined, along the shore, with low trees and bushes. These were completely embowered in running vines, forming columns, arches, and fantastic grottoes.

The sun of the 15th had not risen, when an exclamation of some one called us all out for the first glimpse of Santarem. Surely enough, a white steeple was peeping through the gray mist, bidding us good cheer, for here, at last, we should rest awhile from our labors. The steeple was still some miles ahead, but the spontaneous song of the men, and the hearty pulls at the paddle, told us that these miles would be very short.

Crossing to the southern side, we soon entered the current of the Tapajos. This river is often called the Préto, or Black, from the color of its waters; and, for a long distance, its deep black runs side by side with the yellow of the Amazon, as though this king of rivers disdained the contribution of so insignificant and dingy a tributary. And yet, the Tapajos is a mighty stream. The shore was deeply indented by successive grassy bays, with open lagoons in their centers, about the margins of which various water-fowl were feeding. Most conspicuous in such places is, always, the Great White Egret, Ardea alba, who raises his long neck above the grass as the suspicious object approaches. With an intuitive perception of the range of a fowling-piece, he either quietly resumes his feeding, or deliberately removes to some spot near by, where he knows he is beyond harm. The Heron is sometimes spoken of as a melancholy bird, but whether stalking over the meadows, or perched upon the green bush, he seems to me one of the most beautiful, graceful beings in nature. The Lady of the Waters, a name elsewhere given to a single species, might, without flattery, be bestowed upon the whole.

The trees beyond these bays were many of them in full bloom, some covered with glories of golden yellow; others, of bright blue; and others still, of pure white. Many had lost their leaves, and presented somber Autumn in the embrace of joyous Spring; thus tempering the sadness which irresistibly steals over one when witnessing nature's decay, with the joy that lightens every feeling, when witnessing her renovation.

Leaving these pretty spots, low trees covered the shore, and in their branches, we noticed many new and beautiful birds, that made us long for a montaria.

When near the town, part of our company left the galliota, and walked up along the beach. Our letters were to Captain Hislop, an old Scotch settler, and directly on the bank of the river, at the nearer end of the town, we found his house. The old gentleman received us as was usual, placing his house at once "a suas ordens," and making us feel entirely at home. We walked out, before dinner, to show our passports to the proper officers, although we understood this to be rather matter of compliment than of necessity, as formerly. Not finding the officers, we made several other calls, the most agreeable of which was to Senhor Louis, a French baker, and a genuine Frenchman. He was passionately fond of sporting, and although he had been for several days unable to attend his business from illness, he at once offered to disclose to us the hiding places of the birds, and to be at our disposal, from sunrise to sunset, as long as we should stay.

After our galliota habits, it seemed odd enough to sit once more at a civilized table; but that feeling was soon absorbed in astonishment at Santarem beef, so tender, so fat, so eatable. How could we ever return to the starved subjects of Pará market?

The captain had been a navigator upon all these rivers, and particularly the Tapajos; having ascended to Cuyabá, far amongst its head waters. At Santarem, the Tapajos is about one mile and a half wide, at high water. Above, it greatly widens, and, for several days' journey, is bordered by plantations of cacao. At about twelve days' journey, or not far from two hundred and fifty miles, the mountains appear, and the banks are uneven, and of great beauty. The region thence above, is a rich mineral region, and rare birds, animals, and flowers are calling loudly for some adventurous naturalist, who shall give them immortality. Here are found the Hyacinthine Macaws, M. hyacinthinus, and the Trumpeters, Psophia crepitans. At certain points, the navigation is obstructed by rapids, and to pass these, the canoes are unloaded and dragged over the land. The journey from Pará to Cuyabá requires about five months, owing to the absence of regular winds, and the swiftness of the current. Canoes occasionally come down, bringing little except gold, and in returning, they carry principally salt and guaraná, a substance from which a drink is prepared. At a distance of several hundred miles above Santarem, is a large settlement of Indians, and from them, come the feather dresses seen sometimes in Pará. These are worn by the Tauçhas. A cap, tightly fitting the head, is woven of wild cotton, and this is covered with the smaller feathers of macaws. To this is attached a gaudy cape reaching far down the back, and formed by the long tail feathers of the same birds, of which they also make scepters that are borne in the hand. Besides these, are pieces for the shoulders, elbows, wrists, waist, neck and knees; and often, a richly worked sash is thrown round the body. These dresses are the result of prodigious labor, and far surpass, in richness and effect, those sometimes brought from the South Sea Islands.

From the Tapajos Indians, come also, the embalmed heads, frequently seen at Pará. These are the heads of enemies killed in war, and retain wonderfully their natural appearance. The hair is well preserved, and the eye-sockets are filled with clay, and painted. The Indians are said to guard these heads with great care, being obliged, by some superstition, to carry them upon any important expedition, and even when clearing ground for a new sitio. In this case, the head, stuck upon a pole in one corner of the field, watches benignly the proceedings, and may be supposed to distil over the whole a shower of blessings.

The river, below the falls, is not subject to fever and ague; and above, only at some seasons.

Santarem is the second town to Pará, in size, upon the Amazon, and has every facility, from its situation, for an extensive trade with the interior. It is in the centre of the cacao region, and retains almost entire control of that article. Vast quantities of castanha nuts also arrive at its wharves from the interior. The campos in the vicinity support large herds of fat cattle, in every way superior to those of Marajo; and were steamboats plying upon the river, Santarem beef would be in great demand at Pará. Its population is about four thousand. It stands upon ground inclining back from the river. Its streets are regular, and the houses pleasant looking, usually but of one story, and built as in Pará. It contains a very pretty church, above which tower two steeples. The fort is very conspicuous, standing upon a high point, at the lower end of the town, and commanding the river.

The morning after our arrival, we called upon the commandant and the chief of police. Both were gentlemanly, educated men; and, very kindly, expressed themselves happy to do us any favor, or assist us in any way. At one of these houses, was a very curious species of monkey; being longhaired, gray in color, and sporting an enormous pair of white whiskers.

Another monkey, of a larger species, shaggy, with black hair, was given us as a present. This, we left until our return.

In the vicinity of Santarem, the scarcity of laborers is most severely felt; slaves being few, and Indians not only being difficult to catch, but slippery when caught. We suspected some persons of tampering with our men, and therefore, judged it better to proceed at once, although we had intended to remain several days. Our suspicions proved true, for upon leaving, two of the buys were determined to remain behind, and were only prevented from so doing by our summoning an officer, and the threat of the calaboose. A detention in the calaboose, would in itself be slight; but when it involves, at least, three hundred lashes from the cat, a most detestable animal to the Indian, it becomes something to be considered. Desertion is so common, and so annoying, that it receives no mercy from the authorities.

Leaving Santarem, we crossed to an igaripé, leading into the Amazon. Seen from this distance, the town presents a fine appearance, to which the irregular hills in the background much contribute. The highest of these hills approaches pretty nearly our idea of a mountain. It is of pyramidal form, and is known by the name of Irirá. The igaripé was narrow; lined, upon one side, by sitios, upon the other, by an open campo. While coasting along this, one of the boys, who had attempted desertion, threw himself on the cabin top, in a fit of sulks, and commenced talking impudently with the pilot. A—— told him to take a paddle, which he refused; and, quicker than thought, he found himself overboard, and swimming against the current. He roared lustily for help; and after a few moments, we drew up by the grass, and allowed him to climb in, considerably humbled, and ready enough to take a puddle. This had a good effect upon all; and the alacrity with which they afterwards pulled, was quite surprising.

Chapter XII

***THE AMAZON THUS FAR—A CACAO SITIO—
POLITENESS—RUNAWAYS—GROWING OF CACAO—AN
ALLIGATOR—HIGH BANK—DESERTED SITIO—
KINGFISHERS—ROMANÇAS—WATER BIRDS—ARRIVE
AT OBIDOS—RIO DES TROMBETAS—INCIDENTS UPON
LEAVING—MANNER OF ASCENDING THE RIVER—
SHELLS—STOP AT A SITIO— HIGH BLUFF—WATER
PLANTS—CAPITAN DES TRABALHADORES—ARRIVE AT
VILLA NOVA—FIESTA OF ST. JUAN—WATER SCENE—A
VILLA NOVA HOUSE—TURTLES—STROLL IN THE
WOODS—LAKES***

The river, above the junction of the Tapajos, was sensibly narrower. Between Garupá and Santarem, its width had averaged from eight to twelve, and sometimes fifteen miles. From the mouth of the river to Sanlarem, a distance of six hundred miles, twelve hundred islands are sown broad-cast over the water; many of large size, and but few very small. These have been accurately surveyed, and their places laid down upon charts, by the officers of a French brig of war, within a few years. Owing to this multitude of islands, we rarely had the opportunity of distinguishing the northern shore.

The waters now were decreasing, having fallen between one and two feet. Their annual subsidence, at Santarem, is twenty-five feet; and they do not reach that point, until late in December. At that time, the tides are observable for a distance of several hundred miles, above the Tapajos. Even at the height of water, they cause a slight flowing and ebbing at Santarem.

We had been advised, that the cárapanás were more blood-thirsty above the Tapajos; and our first night's experience, made us tremble for the future.

Early in the morning, June 17th, we drew up by a cacao sitio. The only residents here, were four women; two, rather *passé*, and the others, pretty, as Indians girls almost always are. They were seated upon the ground, in front of the house, engaged in plaiting palm leaves: and to our salutation of "muito bem dias," or "very good morning," and "liceincia, Senhoras," or, "permission to land, ladies," they answered courteously, and as we desired. This was rather more agreeable, than an affected shyness, a scudding into the house, and peeping at us through the cracks, as would have been our reception in some other countries I wot of. Politeness is one of the cardinal virtues in Brazil; and high or low, whites, blacks, or Indians, are equally under its influence. One never passes another, without a touch of the hat and a salutation, either, good morning or afternoon; or more likely still, "viva Senhor," "long life, sir:" and frequently, when we have been rambling in the fields, a passing stranger has called out to us a greeting from a distance, that might readily have excused the formality. An affirmative or negative, even between two negroes, is "si, Senhor," or "nao, Senhor." Two acquaintances, who may meet the next hour, part with "ate logo," or "until soon," "ate manhaã," "until to-morrow." When friends meet, after an absence, they rush into each other's arms; arid a parting is often with tears. "Pussa bem, se Deos quiere," "may you go happily, God willing," is the last salutation to even a transient visitor, as he pushes from the shore; and very often, one discovers, that the unostentatious kindness of his entertainer hats preceded him, even into the boat.

But to return to our ladies. A distribution of cashaça and cigars, quite completed our good understanding; and with the more particularly interesting ones, the popularity of the universal Yankee nation certainly suffered no diminution. They understood the arts of the cuisine too, and assisted us mightily in the preparation of our viands. As a parting gift, they sent on board a jar of fresh cacao wine, the expressed juice of the pulp which envelops the seed, a drink delightfully acid, and refreshing.

While here, our two boys embraced the opportunity to run away, leaving all their traps behind them. It was embarrassing, but there was no remedy, and we consoled ourselves with the suggestion, that after all, they were lazy fellows, not worth having.

We were now in the great cacao region, which, for an extent of several hundred square miles, borders the river. The cacao trees are low, not rising above fifteen or twenty feet, and are distinguishable from a distance by the yellowish green of their leaves, so different from aught else around them. They are planted at intervals of about twelve feet, and, at first, are protected from the sun's fierceness by banana palms, which, with their broad leaves form a complete shelter. Three years after planting, the trees yield, and thereafter require little attention, or rather, receive not any. From an idea that the sun is injurious to the berry, the tree tops are suffered to mat together, until the whole becomes dense as thatch work. The sun never penetrates this, and the ground below is constantly wet. The trunk of the tree grows irregularly, without beauty, although perhaps, by careful training, it might be made as graceful as an apple tree. The leaf is thin, much resembling our beech, excepting that it is smooth-edged. The flower is very small, and the berry grows directly from the trunk or branches. It is eight inches in length, five in diameter, and shaped much like a rounded double cone. When ripe, it turns from light green to a deep yellow, and at that time, ornaments the tree finely. Within the berry, is a white, acid pulp, and, embedded in this, are from thirty to forty seeds, an inch in length, narrow, and flat. These seeds are the cacao of commerce, When the berries are ripe, they are collected into great piles near the house, are cut open with a tresádo, and the seeds, squeezed carelessly from the pulp, are spread upon mats to dry in the sun. Before being half dried, they are loaded into canoes in bulk, and transmitted to Pará. Some of these vessels will carry four thousand arrobas, of thirty-two pounds each, and as if such a bulk of damp produce would not sufficiently spoil itself by its own steaming, during a twenty days' voyage, the captains are in the habit of throwing upon it great quantities of water, to prevent its loss of weight. As might be expected, when arrived at Pará, it is little more than a heap of mould, and it is then little wonder that Pará cacao is considered the most inferior in foreign markets. Cacao is very little drank throughout the province, and, in the city, we never saw it except at the cafés. It is a delicious drink, when properly prepared, and one soon loses relish for that nasty compound, known in the States as chocolate, whose main ingredients are damaged rice, and soap fat. The cacao trees yield two crops annually, and excepting in harvest time, the proprietors have nothing to do but lounge in their hammocks. Most of these people are in debt to traders in Santarem, who trust them to an unlimited extent, taking a lien upon their crops. Sometimes the plantations are of vast extent, and one can walk for miles along the river, from one to the other, as freely as through an orchard. No doubt, a scientific cultivator could make the raising of cacao very profitable, and elevate its quality to that of Guyaquil.

 Towards evening, a little alligator was seen upon a log near shore, and we made for him silently, hoping for a novel sport. One of the men struck him over the head with a pole, but his casque protected him, and plumping into the water, we saw him no more.

The morning of the 18th, found us boiling our kettle under a high clay bank, which was thoroughly perforated by the holes of kingfishers, who, great and small, were flying back and forth, uttering their harsh, rapid notes, and excessively alarmed at the curiosity with which we inspected their labors. We tried hard to discover some eggs, but the holes extended into the bank several feet, and we were rather afraid that same ugly snake might resent our intrusion. Various sorts of hornets, bees, and ants, had also their habitat in the same bank, and so completely had they made use of what space the birds had left them, that the broken clay resembled the bored wood that we sometimes observed in the river below. This clay was of sufficient fineness to be used as paint, and, in color, was yellow and red. When fairly exposed to the sun, it seemed rapidly hardening into stone.

Upon the hill were two houses, one neatly plastered, the other of rough mud, with a thatched roof. Both were deserted, and evidently had been for a long time. Traces of former cultivation were every where in the vicinity, lime and orange trees being in abundance, and the vines of the juramu, a sort of squash, running over every thing. No one knew to whom this had formerly belonged; but probably, to some sufferer by the revolution. Near by the houses, we observed a number of new flowers, one of which was a large white convolvolus, that thereafter we frequently saw upon the shore.

During the morning, we sailed some miles under a bank of one hundred feet in height, usually entirely wooded to the water's edge. But wherever the sliding earth had left exposed a cliff, it was drilled, by the kingfishers to such a degree, that we often counted a dozen holes within a square yard. It seemed to be the general breeding place for all the varieties of this family from hundreds of miles below.

We saw many fine looking houses, and large plantations upon tile hill, and the table land seemed to run back a long distance. Here the fortunate proprietors lived, beyond reach of cárapanás, a most enviable superiority.

The river took a long sweep to the north, describing nearly two-thirds of a circle, and indented by small bays. In these, the water was almost always still, and often flowed back. These latter aids to poor travelers are called romanças, and the prospect of one ahead was exceedingly comfortable. Great quantities of grass are caught in these romanças, and spend a great part of their natural lives in moving, with a discouraging motion, now up, now down, as wind or current proves stronger.

About noon, we passed the outlet of a large lake, or rather of what seemed to be a wide expansion of the waters of the river, between a long island and the southern shore. Here were numerous fishing canoes, and hundreds of terns were flying about as though they, too, considered this good fishing ground. There were also many of the small duck, called the Maraca. Both these varieties of birds were seen in large flocks, wherever logs, projecting from the water, allowed their gathering, and often, hundreds were floating down upon some vagrant cedar. The fields of grass were now a constant feature, and often lined the shore to such an extent as rendered landing impracticable.

Our route, upon the 19th, was extremely uninteresting, passing nothing but cacao trees, whose monotonous sameness was terribly tiresome. By three o'clock, we had arrived at Obidos. Two high hills had, for some hours, indicated the position of the town, but so concealed it, that we were unable to distinguish more than two or three houses, until we were close upon it. In crossing the current, for Obidos is upon the northern side, our galliota was furiously tossed about, and carried some miles below. The main channel of the Amazon is here contracted into a space of not more than a mile and a half, and dashing through this narrow passage, the waters boil and foam like some great whirlpool. The depth of the channel had never been ascertained until the French survey, when it was measured as one hundred fathoms, or six hundred feet. The position of Obidos is very fine, thus commanding the river, and being also at the mouth of a large tributary, the Rio de Trombetas. It was upon this river, that Orellana placed his nation of Amazons. The friar, who accompanied him, affirmed, that they had fought their way through a tribe of Indians, who were commanded by a deputation of these warlike ladies in person, and described them as tall, and of a white complexion, wearing their luxuriant hair in plaits about the head. Their only dress was a cincture, and they were armed with bows and arrows. Expeditions have, at different times, been sent to explore the Trombetas, but, from one cause or another, have failed; and numerous accounts are credited of single adventurers, who have lost their lives by the cannibal, upon in banks. But, no doubt, the country, through which the river passes, is well worthy exploration, rich in soil and productions, if not in minerals.

Odidos contains, perhaps, one thousand people, and is built in the customary, orthodox manner of the country. It has considerable trade, if we might judge by the number of its stores, and the good assortments therein contained.

We walked about, visiting one and another, until evening, the observed of all observers. It was not often that so many foreigners perambulated one of these towns together, and every one seemed disposed to gaze, as though the opportunity occurred but once in a lifetime.

It was delightful to see a horse once more, for we had not enjoyed that privilege since leaving Pará. Here also was an Indian Hog, or Peccary, running about the streets, and appearing in his motions and habits, as any other hog.

We were under some apprehension of losing more of our crew, and made preparations for leaving immediately. But considering that our circumstances afforded as fair an excuse as those of our neighbors, we offered the pilot a patac for every "good and able-bodied seaman" that he would enlist. This put him upon his mettle, and, as soon as dark set in, he was up and down the beach, surrounded by several acquaintances whom he had picked up, and eloquently depicting the advantages of regular wages, and rations of coffee and cashaça.

Eloquence is "the art of persuasion," and our pilot was a gifted man; for, in a short time, he had engaged five men, and more were waiting his approaches. But we had now our complement, and, by midnight, were under way, the whole crew in a most glorious state of jollification. The old Tauçha, quiet old man as he usually was, lay sprawled upon the top, sputtering unknown tongues, and singing with vigor enough to arouse the garrison. In one of his activities, he rolled off, and this seemed to freshen him a little, for after we had given him a lift out of the shallow water into which he had kicked and plunged himself, he became comparatively decent. The men, most of them, rowed with a fervor quite delightful, and we had crossed the river, and were proceeding rapidly, when, souse went another, dead drunk, from the cabin top. Strange that cold water should have had so instantaneous an effect, but, log-like as he was, he revived at once and pulled for the grass, from which we took him in. It was scarcely worth while to advance in this manner, so to prevent further mishaps, we ran the bow into the grass, and waited a more propitious morning.

The next morning, the men were in more sensible order, and a pull of a few hours, before breakfast, made them once more themselves. The Tauçha was as kingly as ever, and placid as a summer's morning. It was amusing to hear him joke with the pilot, about the man who fell overboard, and as often as he thought of it, his fat sides would shake with inaudible laughter. Evidently, he had entirely forgotten his own bad plight.

The wind was fair, and we sped rapidly. We passed a long, low flat, covered with grass, interesting to us as these campos always were, from the great variety of birds that congregated upon them. Here we first observed a small bird of the Tody species, with head and shoulders of white, the body being black. It was the T. leucocephalus, and was usually seen in the grass, rather than on bushes or trees. Here, also, were many Red-throated Tanagers, T. gularis, a very common species, but striking, from its contrasted colon of red, white, and black. Beyond this campo, long lines of willow trees skirted the shore, their leaves mostly fallen; and the whole tableau looked any other than a tropical one. We passed one of the arms of the river. Heavy waves dashed over our sides, and we felt what a slight protection our overloaded craft would be, if overtaken by one of the squalls, so common at this season, but which we, fortunately, had not yet experienced.

We had now left the cacao plantations, and again welcomed the wild beauty of the forest border, where the birds might sing, and the monkeys gambol for our amusement, as merrily as though white men had never passed these waters.

Towards night, we saw a large vessel, which was breasting the current in an altogether novel way to us. A montaria went ahead, dragging a long rope, one end of which was fastened to the bow. This rope was tied to some convenient object on shore, and hand over hand, those in the vessel pulled her up; when the same process was repeated. In this manner, she advanced about one mile an hour, and this is the custom with all large craft, when wind does not favor.

During the night, the breeze died away, and for several days thereafter, was, if blowing at all, dead ahead, so that our progress was discouragingly slow. Upon the 21st, the heat was most oppressive, and, to add to our discomfort, the current ran so furiously, that the utmost exertions of the men, could, at times, scarcely propel the boat. About noon, we passed a large house, upon a small bluff, adjoining which was a chapel and a number of small cottages. Altogether, it was the finest establishment that we had seen, since entering the Amazon. Not far above, we stopped to breathe a while at a sitio, and in wandering about the mandioca plantation, we discovered a number of shells, but of similar varieties to those found below. Growing upon this place, were pepper plants, in abundance, and the Indians had soon stripped them of their berries. One could not but wonder what the stomachs of these men were lined with, when, with every mouthful of farinha, they threw in a fiery red pepper, the very sight of which was almost enough to season a dinner. Yet, the whites also acquire this habit, and eat the article with as much relish as the Indians.

Upon the 22^d, the course of the river was very tortuous, so that, at no time, could we discover the channel far in advance. High lands towards Villa Nova began to skirt the horizon to the westward. We gathered a new variety of cactus, running over the tree-tops like a vine; and a lofty tree which we passed, was draped with the nests of the large Crested Troopial, Cassicus cristatus, three feet in length. There is another variety, more common below, the Cassicus viridis, or Jacú, and usually encountered in the deep forest. Both these species are nearly the size of crows. We saw, during the morning, an unusual number of our favorite Thrush, D. vociferans. Wherever a grassy spot was seen, his song was sure to come trilling out of it, and with very little shyness, he would allow us a fair sight of his beautyship, as he sat perched upon some tall spear, or chased his mate sportingly through his mimic forest. Just before dark, we arrived at the house of a Villa Nova padre. He was not at home, but a number of Indian women seemed to be the managers, and from them, we obtained a pair of Tambaki, a fish much esteemed upon this part of the river, and a turtle. These turtles were now ascending the river to their breeding places upon the upper tributaries, and upon several occasions we had observed them floating upon the water, near our boat.

Early upon the 23^d, we passed a high bluff, which marks the Upper from the Lower Amazon. Below, we had been in the district of Pará; now, we had entered that of the Rio Negro.

We saw increasing quantities of a very pretty water plant, whose flowers were blue and white, and about the basis of whose leaf-stems were spongy expansions, always filled with air—natural swimming corks.

The sun was just setting, as we drew up at the sitio of the Capitan des Trabalhadores, to whom we had letters from Doctor Costa, desiring him to arrange men for our further advance. He promised to go to town in the morning, and filling one of our lockers from his orange trees, we proceeded on. Villa Nova is not upon so high land as some of the towns below, and is not conspicuous from a distance. But its situation is marked by an opposite island, the upper point of which extends two or three miles beyond the town. This was watched by many eager eyes, for it was the eve of the Festa of St. Juan, one of their most popular of saints; and our men, if possible, were more anxious than we, and strained every nerve to arrive in time for the evening's festivities. With such a will, it was not long before the roaring of the muskets, deputized as cannon, and the bright light of bonfires, burst upon us. Suddenly, the whole illuminated town was before us, bonfires glaring before every door, and an especially large one at the upper end, where the Delegarde resided.

We came in among a crowd of montarias and large canoes, mostly filled with women, whom their husbands and fathers had deserted for the more attractive cashaça shops, and who were patiently awaiting the hour of the dança. Upon the bank a procession was passing, the front rank noisy in the plenitude of drums and fifes. Succeeding them were ingeniously preposterous angels; some, overtopped by plumes several feet in length; others, winged with a pair of huge appendages, looking like brown paper kites; and others still, in parti-colored gauds, suggestive of scape-angels from Pandemonium. Behind these loitered the tag, rag, and bobtail, or the black, red, and yellow, in the most orthodox Tammany style.

Some of our party went on shore to look up old acquaintances. I remained on board, preferring to make observations by daylight. It was late before the noise in the town subsided, what with muskets and rockets, singing and fiddling; so late, that I must have been dreaming hours before; but the first thing that awoke me in the morning, was a splashing, and laughing, and screaming all around the galliota, where the sex, *par excellence*, was washing away the fatigues of the dance in a manner to rival a school of mermaids. And these Indian girls, with their long floating hair, and merry laugh, would be no bad representations of that species not found in Cuvier; darting through the surf like born sea-nymphs.

We were invited to the house of Senhor Bentos, a warm-hearted old bachelor, and his little reception room, of perhaps, twelve feet square, was soon festooned with our hammocks. Here we spread ourselves at ease, as if no such vanities as Amazon voyages existed, and waited for the turtle that was undergoing a process in the Senhor's kitchen.

Meanwhile we took the bearings of the Senhor's house, and as it was much like the other buildings of the town, its description will answer for all. Its framework was of rough poles from the forest, and these, within and without, were plastered with brown clay. The floor was of the same material, and the roof was of palm leaves, instead of tiles. From the outer door, a broad hall crossed the house, and this, being used as a dining-room, was occupied by a long table, upon either side of which was a four-legged bench. From the hall, upon each side, opened a small chamber, one used as the sleeping apartment of the family, and the other, in which we were swinging, the Senhor's especial parlor, or bedroom, as the case might be. In this was a large window, closed entirely by a shutter. The whole structure, to our ideas, was rather comfortless; but, under the equator, that is of small consequence, and sufficient comfort is centered in a hammock, to atone for its absence in every thing else. Back of the house was a covered kitchen, and around this was a yard well stocked with poultry, and shaded by orange trees.

The dinner came off in good style, and turtle in every variety of preparation, from the soup to the roasted in the shell, tempted us. It was the first time we had seen the turtle of the Amazon, and in our enthusiasm, we pronounced it equal to the very best of varieties seen at the North, nor wondered, that at civic dinners, aldermen must, perforce, make gluttons of themselves.

After dinner I strolled into the woods back of the town, and soon discovered a delightful path, where a coach and four might have driven.

At no great distance was a burying-ground, marked by a lofty cross, but as yet, apparently, without a grave. As I loitered along, picking here and there a flower, or startling the lizard from his afternoon nap, a number of Indians in their gala dresses, the women with bright flowers in their hair, passed, all greeting me with the musical "viva," or "como esta, Senhor."

Towards evening, the festivities of the day being over, one after another, the canoes about the galliota pushed off, leaving the town almost deserted. Some of our men endeavored to take French leave of us, for which they enjoyed the night in the calaboose.

There were some cattle about Villa Nova, and next morning, the 25th, was rendered memorable by the acquisition of a goodly quantity of milk, the first real cow's milk that we had seen, since

New York milkmen used to disturb our early dreaming. And even this good milk tasted all the more natural for a dash of water.

We were very desirous to see the lake that lies about a mile in the rear of the town, but were prevented by the weather. In this vicinity, a chain of lakes extends along the river, upon both shores, and far into the interior. This lake region is generally high land, and uninfested by cárapanás. Multitudes of Indians are scattered over it, obtaining an easy subsistence from the vast numbers of Periecu, and other fish, which frequent the lakes. At this season also, turtle resort to the same places, and were beginning to be taken in great numbers.

Since leaving Pará, our movements had been pretty much restricted to the galliota, for want of a montaria in which we might visit the shore at our inclination. At Villa Nova, we were fortunate enough to purchase one convenient for our purposes, and now anticipated a great increase to our means of amusement. And yet, our time heretofore had passed most pleasantly. The skies had favored, and those of us who were inclined, spent our days upon the cabin top, shielded from the boards by a comfortable rug, and shaded from the sun, if need were, by umbrellas. But the sun's heat was rarely inconvenient, and tempered by fresh breezes. Coasting close in shore, there was always matter for amusement; in the morning and evening, multitudes of birds, and, at all hours, enchanting forests or beautiful flowers. At night, we preferred the open air to the confinement of the cabin, and never wearied in admiring the magnificence of the skies, or in tracing the fantastic shapes that were mapped out upon them in a profusion inconceivable to those who are only acquainted with the skies of the northern hemisphere. I have alluded to this before; but so interesting a phenomenon deserves further notice. This increased brilliance of the tropical skies is owing to the purity of the atmosphere, which is absolutely free from those obscuring, murky vapors, that deaden light in other latitudes. The sky itself is of the intensest blue, and the moon seems of increased size and kindlier effulgence. For one star at the North, myriads look down with a calm, clear light, and great part of the vault is as inexplicable as the milky-way. Most beautiful in appearance, and interesting from association, is the Southern Cross, corresponding with the Great Bear of the North. This constellation is of four stars, of superior brilliance, arranged in the form of an oblique angled cross. Just above these, and seeming to form part of the same constellation, is the Centaur. Orion is in all his glory, and the Scorpion trails his length, most easily recognized of all. All the other zodiacal clusters are conspicuous, and a kindred boat we do not care to name.

As the sun always set about six o'clock, we had long evenings, and it was our custom to gather upon the cabin, and while away the hours in singing all the psalms, and hymns, and social songs, that memory could suggest. Old Amazon was never so startled before; and along his banks, the echoes of Old Hundred and Lucy Long may be traveling still.

The cárapanás had not been so troublesome as we had feared, and we had often avoided all their intimacies by tying to some tree removed from shore, or by favor of the fresh breezes.

Chapter XIII

***LEAVE VILLA NOVA—OUR MANNER OF LIVING—
SHELLS—JACAMARS—PAROQUETS—MONKEYS—
SCORPION—ENTER AN IGARIPÉ—A DESERTED SITIO—
WILD DUCK—SCARLET TANAGERS—A DESERTED
SITIO—TOBACCO—SHELLS—A COLONY OF MONKEYS—
A TURTLE'S REVENGE—IMMENSE TREES—ALBINO
MONKEY—A SELF-CAUGHT FISH—PORPOISES—
CURASSOWS AND NESTS—A TURTLE FEAST—
SQUIRREL—WILD INDIANS—WHITE HERONS—
SHELLS—UMBRELLA CHATTERER—CROSS TO THE
NORTHERN SHORE—PERIECU AND TAMBAKI—ARRIVE
AT SERPA—SR. MANUEL JOCHIN—AN INDIAN DANCE***

The sun of the 26th June was just relighting the water as we left Villa Nova. Continuing on a few miles, we stopped in the woods to breakfast. Our friends had loaded us with provisions of fish, fowl, and turtle, and this morning's picnic was peculiarly delightful after the Sparian fare of the last fortnight. And here, perhaps, a description of our doings at these breakfast hours may not be without interest, to those who care to know the romance of a voyager's life. Landing at a convenient spot, the first point was to clear a space sufficient for operations, and this was speedily effected by some of the Indians, with their tresádos. Others wandered about collecting materials, wherewith to make a blaze, and there was rarely difficulty in finding an abundance of such. The flint and steel were put in requisition, and soon all was ready. Some of the party cut off strips of fish, washing it to extract the saltiness; others cut sticks of proper length, into the cleft end of which they fastened the fish. These were then stuck in the ground, inclining over the fire, and one of the men was always stationed near to give it the requisite turning. One of the Indians was the particular attendant upon the cabin, receiving sundry perquisites for his services; and upon him devolved the care of our tea-kettle. Above the fire, a cross-bar was supported by a forked stick at either end, and on this, the boiling was accomplished in the most civilized style. The coffee bag was all in waiting, a flannel affair, which, whilom, had done duty as a shirt sleeve; and into this was put about two tea-cups of coffee. The boiling water was poured in, and our wash-bowl, *washed*, received the beverage, fragrant and strong. A quart was the allowance for each, and this, properly attempered by sugar, and unspoiled by milk, was our greatest luxury. As to the more substantial moiety of breakfast, the fish, rank and tough, we stood not upon ceremony, but pulling it in pieces with our fingers, and slightly dipping it in a nicely prepared mixture of oil and vinegar, we thereafter received it as became hungry men.

At times, our fare was varied by the articles obtained at some sitio, but this was the general rule. Two of us had left the North dyspeptics. Sufficient was cooked in the morning to serve us through the day, and therefore we usually made but one stoppage.

About the roots of the trees at this place, we found a beautiful variety of shell, the Bulimus papyracea, in considerable numbers, and here also we obtained a richly plumaged Jacamar, the Galbula viridis. This species we afterwards frequently encountered both in the forest and about plantations. There was one other species common at Pará, but less beautiful, the G. paradisea. These birds resembled the humming birds so much in shape, that the people of the country, universally call them "beijar flor grande," or the great kiss-flower. Their lustrous plumage assists the deception. They live upon insects, which they are very expert at catching, with their long, slender bills.

During the morning, we tested the capabilities of our new montaria, and starting in advance of the galliota, found fine sporting, principally among the paroquets and herons. The former family of birds had not been very plentiful since leaving Gurupá, near which place they had collected in vast flocks, from a large extent of country, for the breeding season. But now again, we were in the vicinity of some other haunt, and they were scarcely ever out of sight, or hearing. Their notes were not extremely agreeable, being little more than a shrill chatter, but for beauty of appearance, and motion, when clustered around some tree top, busily engaged in stripping off the berries, they were great favorites with us. There is no enumerating the different varieties we observed, some little larger than canaries, others approximating in size to their cousins, the parrots. In general, their plumage was green, but they differed in their markings, the green being beautified by various shades of yellow, of blue, and of pink, or roseate.

Our advance was not very great, for the wind did not favor us, and all day we were coasting about the greater part of a circle, with the situation of Villa Nova scarcely ever out of sight. We observed very few houses; the land was low, and palms again were numerous. Frequently, turning some point, we came upon little squads of monkeys, who scampered in terrible alarm, at the first glimpse of us. Excepting on these sudden surprisals, it always was exceedingly difficult to catch a sight of these animals. Even when one is positive that some of them are in his immediate vicinity, none but the keen and practiced eyes of an Indian can discover their retreat. For any other than an Indian, therefore, to venture upon a monkey hunt is almost useless, and they only succeed by stripping off their clothes, and creeping cat-like among the bushes, or patiently waiting their opportunity in some concealment.

From a passing montaria, we purchased a fish weighing about fifteen pounds, for four vintens, or four cents. We had noticed that most of the fish that we had seen had broad, flat heads, and corresponding mouths; and this specimen showed us the utility of such a shovel-like apparatus; for, in his stomach, were at least a quart of crabs, as good as new, which he had gathered from the bottom of the river. When the refuse parts of this fish were thrown into the water near shore, they attracted great numbers of a small white fish, which strongly resembled eels in their habits, burying themselves in the mud at any attempt made to catch them. We succeeded in obtaining as many as we wanted of these, at another time, by letting down a basket in which was a bait of meat. Upon pulling this out, half a dozen of these fish were always inside. The Indians would not eat them, but pronounced them "devils" of fishes.

While clearing out one of the lockers this afternoon, we started a brood of scorpions, a kind of reptile more formidable in ancient story that in modern reality. Still, I should prefer not to be stung by one of them. We saw them frequently in different parts of the country, and occasionally, several inches in length. They abound in all canoes and vessels, and once, as I opened a letter, brought from Pará in one of these craft, a nice little specimen dropped from the folds.

Soon after dark, a tremendous storm of wind and rain set in, which twice broke us from our moorings, and deluged the cabin. Rain had no sooner ceased, than swarms of cárapanás hurried to our attack, and for the remainder of the night, sleep was out of the question.

The river, upon the morning of the 27th, made a wide bend to the northward, around an immense inland; and to shorten the distance, we took the smaller channel, which, in narrowness, resembled an igaripé. Here, we again heard the Guaribas, who almost deafened us by their howling.

Towards night, we stopped a few moments at a deserted plantation. The house was in ruins; but the fruit trees, and the garden, were still productive. In a trice, the whole were stripped, as though a party of licensed foragers had chanced that way; and plantains, squashes, sugar-cane, and peppers, were handed into our boat.

Proceeding, we passed a clump of grass, where a duck was setting upon her nest. Starting off, she fluttered along the water, as if badly wounded, and some one sprung to follow her in the montaria; but, before that could be got ready, she had fluttered beyond harm's reach, and then had vigorously flown out of sight

During the day, we had seen a number of birds new to us; but most attractive of all, was a Scarlet Tanager, the Rhamphopis nigri gularis (Swain), or Black-masked, whose brilliant metallic scarlet and black livery, was like a jewel in the sunlight. We had seen nothing comparable to it upon the river. These birds were always seen about low bushes, by the water-side, catching their favorite insects, and uttering a slight note, or whistle, but no song.

The morning of the 28th, found us still in the igaripé, which had become extremely narrow. The shore, upon one side, was two feet above the water; upon the other, it was overflowed. This contrast is observable upon the main stream, and between almost all the islands; high banks being generally opposed by low swamps.

By ten o'clock, we had re-entered the river; and stopped at a sitio, directly upon the point of the island, to prepare our breakfast. This plantation evidently belonged to a more industrious planter, than was usual. There was a fine orchard of young cacao trees, and a large field of tobacco, nicety cleared of weeds. The tobacco, grown in thin district, is of superior quality, and vastly preferred to any American tobacco imported. When put up for use, it is in long, slender rolls, wound about with rattan, and is cut off by the foot. Sometimes, these rolls are ornamented by the Indians, with feathers. All persons, men and women, use tobacco in smoking; and, for this purpose, have pipes of clay, the stems of which are ornamented reeds, three or four feet in length. In the towns, very good cigars are made. We never observed the practice of chewing the weed, among our Indians; but they were always furnished by us with as regular rations of tobacco, as of cashaça. When pipes were wanting, they made cigarillos of the fine tobacco, wrapped in a paper-like bark, called towaré; and one of these was passed around the deck, each person, even to the little boys, taking two or three puffs in his turn, with which he was content for an hour or two, when the process was repeated.

Wandering about this plantation, we discovered a number of shells of three species; two of which were Helices, and hitherto undescribed. The third was the Achatina octona (Des), and observed at Pará.

The Senhor had a large quantity of fish to sell, and we bartered cloth for enough to last us the remainder of our journey. To show the obstructions to profitable labor, the prices received by this man, is a good illustration. Fish, at Villa Nova, was worth two milrees and a half an arroba; and tobacco, being just then scarce, much more. But, although he might have reached Villa Nova in a few hours, yet the return passage was so difficult, that he preferred to receive one milree an arroba for each, and that in barter. In the same way, we bought of him, for about forty cents, a turtle, weighing, at least, one hundred and twenty-five pounds, which he had lanced the day before. There was a Red and Yellow Macaw, Macrocercus Aracanga, in singularly fine plumage, climbing about the trees by the house; and we longed to possess him, but our boat was too crowded.

Leaving this place, we coasted along the northern bank, and for a long time were passing high cliffs of red clay; sometimes perpendicular and overhanging the water, at others, running far back among the trees, and presenting a beautiful contrast of colors. These banks might well be mistaken for stone, were it not for the tell-tale kingfishers.

Suddenly we came upon a colony of large, bushy-tailed Monkeys, who, to the number of, perhaps, a hundred were gamboling about the tops of a few tall trees. The first glimpse of us, put an end to their sport, and away they scampered, helter-skelter, old ones snatching up young ones, and young and old possessed with but one idea. Those who could, made prodigious leaps into the trees below, catching the branches with their long tails, and swinging out, plunged yet again, and were lost to view. Others scrambled down the trunks, or concealed themselves in forks and crevices; and in far less time than I have taken to describe the scene, not a monkey was visible. We passed on; some bold veteran ventured a whistle, another and another returned it, and shortly, we could see the tree tops bending, and hear the rustling of the leaves, as the whole, troop hastened back to their unfinished games.

Towards evening, the wind freshening, we crossed the channel, and now understood ourselves to be upon the shore of the great Island of Tupirambira, the Tapinamba of early voyagers, which, formed by the outlets of the River Madeira, stretches along many leagues.

During the night, we were awakened by a groaning among the men. One of them had gone down to bale out the hold, and having to do so by the side of the turtle, had thought it would be as well to ascertain upon which end was the animal's head. The first feel was both satisfactory and unfortunate; for turtle, not comprehending the intentions of these inquisitive fingers, seized a thumb in his mouth, and squeezed it, rather gently for a turtle, but still, forcibly enough to hint his displeasure. Had he been one of the denizens of our Yankee ponds, the victimized boy would have had a serious search for his old member; as it was, he was disabled, and we thereafter promoted him to the helm.

Not finding a sitio, we stopped upon the 29th, in a forest of magnificent growth, where the open space allowed a free ramble. The bank was three feet above the water, and the fronting trees and shrubs were densely overrun by a vine, producing a profusion of small white flowers, much resembling the Clematis. Many of the trees here, were of enormous size, and had we measured the girt near the ground, would have given us from forty to fifty feet. This seems wonderful, but the explanation is simple. Ten or fifteen feet above the ground, these trunks are round, and not often more than four or five feet in diameter; but, at about that elevation, set out thin supports diverging in every direction, presenting the appearance of a column, supported by a circle of triangles around its base. Of all these trees, the most conspicuous for beauty, was the mulatto tree, mentioned before, and which grew here in abundance.

To-day we obtained a specimen of the Least Bittern, Ardea exilis, and saw a number of Crested Curassows, or Mutuns, as they are called, but were unable to shoot them. We saw, also, many iguanas, who, at our approach, would drop into the water, from the branches upon which they were feeding. But a greater oddity, was a small monkey, white as snow, and undoubtedly an albino. We drew up to the shore, and endeavored to find his hiding place, but unsuccessfully.

Upon the flowers, this day, we observed great swarms of butterflies, of every size and color. A large one of a rich green, was new to us, and most curious, but the brilliant blue ones, seen so often near Pará, still bore the palm for splendor.

Towards evening, a piece of floating grass passed by us, upon which laid the remains of a fish, about five feet in length. He had thrown himself from the water, and there had died. A great variety of the river fish have this habit of leaping above the surface, and not unfrequently, fall into a passing montaria. Our Indians alleged this as a reason for not sleeping in the montaria, which would have accommodated two or three of them with far more comfort than the galliota, where, part of them slept, slung across the tolda like so many sacks, and the rest along their narrow seats, as they could find room.

Upon the morning of the 30th, we were called out to observe a school of porpoises, that were blowing and leaping all around us. This fish resembles much the sea-porpoise in its motions, and is common from Pará up. Its color is pinkish, upon the belly, and a number of them gamboling about is an exceedingly beautiful sight. They are not eaten, and are valuable only for their oil.

A party went out in the montaria, and returned with a pair of White Herons, A. alba, and in the tree tops, in the vicinity, were a large number of these birds, evidently just commencing their nesting.

As we drew up by the bank for breakfast, a Crested Curassow, or Mutun, Crax alector, flew from the top of a low tree near us, and one of the Indians darted up for her nest. There were two eggs, and tying them in his handkerchief, he brought them down in his teeth. These eggs were much larger than a turkey's egg, white, and granulated all over. The Crested Curasraw is a bird about the size of a small turkey. Its general plumage is black, the belly only being white, and upon its head, is a crest of curled feathers. This species has a yellow bill. It is called the Royal Mutun by the Brazilians, and, in the vicinity of the River Negro, is not uncommon. With several other varieties of its family, it is frequently seen domesticated, and in a graceful and singularly familiar bird in its habits. According to some authors, this bird lays numerous eggs, but each of the three nests which we found, during this day, contained but two, and the Tauçha assured us that this was the complement. The nest was, in every case, about fifteen feet above the ground, and was composed of good-sized sticks, lined with leaves and small pieces of bark.

We determined on the immolation of our monster turtle, and all hands, kettles, and pots were in requisition. About a peck of eggs were taken from her, and reserving these, with the hind quarters, and the parts attached to the lower half of the shell, we turned the remainder over to the Indians, who very soon had every part, even to the entrails, stewing in their earthen vessels. The eggs, mixed with farinha, were very delicious, but in my case, at least, they caused an awful reckoning, and for a long time, I could scarcely think of turtle without a shiver.

Soon after starting, we found two other Mutuns' nests, and as the boy climbed to the last, there was a crash and a fall, and we thought his Indian skill had, for once, deserted him. But the commotion was caused by a pair of iguanas, who, from a good height, had precipitated themselves into the water. The rascals, no doubt, had been calculating on an omelet breakfast. This afternoon, we shot a gray hawk, and on picking him up, we found a large red squirrel, of a species new to us, by his side, upon which he had but just commenced dining. This squirrel had legs and tail greatly disproportioned to his body, and we concluded with an acute theorist, that his ancestry had lived so long among the monkeys, as to have become assimilated.

Upon the morning of July 1st, we stopped at a sitio, where was an extensive plantation of mandioca, and another of cacao, and in the vicinity we shot a number of Jacamars and Tanagers, as well as a squirrel of large size and better proportions than our acquisition of the day before.

Near this place was a sideless shanty, where a party of wild Indians had squatted. There was an old crone, two young girls, and a boy of sixteen, all looking miserably enough. The only articles they seemed to possess, were a couple of hammocks, and a large fish roasting on some coals told how they subsisted. These Indians were of the Muras, the same as our Tauçha, and he went over to have a talk with them. Gipsy like, they often come out in this way, and remain, until some depredation obliges them to decamp. This tribe, in particular, are arrant thieves, and semi-civilization did not seem to have eradicated much of the propensity in those of our party, for several times, we had missed little articles, as knives, which, we had no doubt, were carefully preserved in some of the trunks in the tolda.

All day, the shore continued low, but just above the present height of the river, and a few weeks before, evidently they had been entirely flooded. Of course, there were but few sitios.

Just at night, we came upon an immense flock of herons, roosting in the trees upon a small island. A—— went towards them with the montaria, and brought down enough of them for the morrow's breakfast. The survivors flew round and round in puzzled confusion, then wheeled towards another island, where darkness prevented his following them.

Stopped in the woods, upon the 2^d, and upon the roots of the large trees, we collected a number of shells, the Bulimus piperitus (Sowerby), entirely new to us. There were also many shells, three varieties common throughout the river region, Ampullaria crassa (Swain); Ampullaria scalaris (D'Orbigny), and Ampullaria zonata (Wagner), and usually found just above high-water mark. They crawl up there adventurously and are left by the retiring flood. Occasionally, in these forests, we discovered dead shells of the Achalina flaminea. Here we saw a pair of the Umbrella Chatterers, Cephalopeterus ornatus, among the rarest and most curious of Brazilian birds. They were sitting near together, upon the lower branches of a large tree, and a shot brought down the female. Unfortunately, the gun had been loaded but in one barrel, and before ammunition could be obtained from the boat, the male, who lingered about for some moments, had disappeared. We afterwards obtained a fine male upon the Rio Negro. These birds are of the size of small crows, and the color of their plumage is a glossy blue-black. Upon the head, is a tall crest of slender feathers, whence it derives its name, and upon the breast, of both male and female, is a pendant of feathers, hanging to the length of three inches. They are, like all the Chatterers, fruit eaters. They are pretty common upon an island a few days' sail above the Barra of the Rio Negro, but they are not found any where in that region in such flocks as others of the Chatterer family. The Indian name for these birds is Urumuimbu, and the Tauçha informed us, that they built in trees, and laid white eggs.

During the day, we crossed from one island to another, and at last, were again upon the northern side.

Early the next morning, the 3rd, we were overtaken by a small canoe pulled by eight men, and come of our party were delighted to discover in the proprietor an old acquaintance. After mutual compliments and inquiries, the canoe shot past, and we soon lost sight of her. While we were looking out for a place whereon to build our customary fire, the smoke of some encampment ahead caught our eyes, and directing our course thither, we found our friend of daybreak, nicely settled upon a little clearing which he had made, under the cacao trees of a deserted plantation. He politely made room for us, and sent us coffee from his own boat.

Not long after noon, we stopped at a house, where a number of Indians were collected about a Periecu, which they had just caught. This was the fish whose dried slabs had been our main diet for the last few weeks, and we embraced the opportunity to take a good look at so useful a species. He was about six feet long, with a large head and wide mouth, and his thick scales, large as dollars, were beautifully shaded with flesh color. These fish often attain greater size, and, at certain seasons, are very abundant, especially in the lakes. They are taken with lances, cut into slabs of half an inch thickness, and dried in the sun, after being properly salted. It is as great a blessing to the Province of Pará, as cod or herring to other countries, constituting the main diet of three-fourths of the people. We bought, for eight cents, half this fish, and for six more, a Tambaki, weighing about ten pounds. This is considered the finest fish in this part of the river, and resembles, in shape, the Black Fish of the North.

Not far above this sitio, was the village of Serpa, and a turn of the river presented it to us in all the glory of half a dozen thatched houses. So aristocratic an establishment as our galliota was not to come up without causing a proper excitement, and, one after another, the leisurely villagers made their appearance upon the hill, until a respectable crowd stood waiting to usher us. Hardly had we touched the shore, when a deputation boarded us for the news, and we were forced to spend half an hour in detailing the city values of cacao, and fish, and tobacco, and the hundred other articles of traffic. Indeed, this had been our catechism ever since we entered the river, and as we were profoundly ignorant of the state of the Pará market, we had been obliged to invent a list of prices for the general circulation.

The bank, upon which the village stands, rises abruptly about fifty feet above high water mark, but fortunately, in one point, a broad, natural gully allows easier ascent, and, up this, we made our way. Our principal business in stopping here, was to obtain men, if possible, part of ours being lazy, and part disabled from one cause or another. Moreover, the river current above Serpa flows with a vastly accelerated swiftness, rendering more men almost indispensable. We directed our way to the house of Senhor

Manoel Jochin, the most influential man of the village, although not a public officer. Nor had we far to go, for Serpa has been shorn of its glory, and dilapidation and decay meet one at every turn. The Senhor was sitting at his door, in earnest conversation with the Colonel and the Juiz de Paz, and received us not cavalierly, but as became a cavalier. For Senhor Manoel had been a solder in his day, and, although on the shady side of sixty, still looked a noble representative of those hardy old Brazilians who have spent their lives on the frontiers. We had heard of him below, as the captor of Edoardo, one of the rebel Presidents of the Revolution, and looked upon him with interest. For this exploit he had been offered a high commission in the army, but he preferred living in retirement here.

In the evening, we sat down to turtle and tambaki with the dignitaries before mentioned, and as our style of supper varied somewhat from our former experience, I trust I shall be excused for entering a little more into particulars. By the side of each plate, was a pile of farinha upon the table, and in the centre, stood a large bowl of caldo, or gravy. Upon sitting down, each one, in turn, took up a handful of his farinha, and dropped it into the bowl. This, afterwards, was the general store, from which each helped himself with his own spoon, as he listed. Water was not absolutely interdicted, but it was looked upon with scarcely concealed disapprobation, and its absence was compensated by cashaça. There was no limit to hob-nobbing and toasting, and our jolly Colonel at last concluded with a stentorian song.

The Senhor had been a frequent voyager upon the Madeira, and gave us interesting accounts of his adventures upon that river. What was quite as agreeable, however, was a collection of shells which he had picked up along its shores, and of which he begged our acceptance. One of these was a remarkably large one of the Ampullaria canaliculate (Lam.), which was used as a family cashaça goblet. The others were Hyria avicularis, and Anadonta esula. The valves of the Anadontas had been used as skimmers, in the Senhor's kitchen.

We were told that there was to be a dance, to which our company would be acceptable, particularly if we brought along a few bottles of cushaça. Now an Indian dance was a novelty, and the insinuating invitation worked its effect. Taking each a quart bottle under his arm, we strolled to the scene of action, and were politely ushered into one of the larger houses, where a crowd of men and girls had collected. The room was illuminated by burning wicks of cotton, which were twisted about small sticks, and set into pots of andiroba oil. Around the walls, were benches, upon which sat a score of Indian girls, dressed in white, with the ever accompanying flowers, and vanilla perfume. The men were standing about, in groups, awaiting the commencement of the exercises, and dressed in shirts and trousers. One, distinguished beyond the rest, by a pair of shoes, and a colored handkerchief over his shoulders, was the major domo, and kindly relieved us of our bottles, allowing us to stand ourselves among the others, as we might. A one-sticked dram soon opened the ball, assisted by a wire-stringed guitar, and for a little time, they divinized on their own account, until they were pronounced safe for the evening. Two gentlemen then stepped up to their selected partners, and gracefully intimated a desire for their assistance, which was favorably responded to. The partners stood opposite each other, and carelessly shuffled their feet, each keeping slow time, by the snapping of their fingers. The man advanced, then retreated, now moved to one side, and then to the other. Now approaching close to the fair one, he made a low bow, looking all sorts of expressions, as though he was acting a love pantomime; to which his partner responded by violently snapping her fingers, and shuffling away as for dear life. Away goes the lover two or three yards to the right, profoundly bowing; then as far to the left, and another bow. Getting visibly excited, up again he advances, going through spasmodic operations to get louder snaps from his fingers. The fair inamorata is evidently rising. Around she whirls two or three times; he spins in the opposite direction, and just as he is getting up an attitude of advance, out steps another lady, taking his partner's place. This is paralyzing, but the lover is too polite not to do a little for civility, when some gentleman steps before him, taking the burden from his feet, and leaving him to follow his partner to the well earned seat, where he solaces his feelings by a long pull at the bottle, and then passes it to the lady, who requires sympathy similar in degree and quantity. The dancing continued, with no variation of time or figure, until the cashaça gave out, which was the signal for a breaking up, all who could preserve their equilibrium, escorting their equally fortunate partners, and those who could not, remaining until a little sleep restored their ailing faculties.

Chapter XIV

FOURTH OF JULY AT SERPA—LAKE SARACÁ—AN ACCESSION—PICNIC—AN OPOSSUM—NARROW PASSAGE—SWALLOW-TAILED HAWKS—SITIO OF THE DELEGARDE—RIVER MADEIRA—VILLAGE OF OUR TAUÇHA— APPEARANCE OF HIS PARTY ON ARRIVING AT HOME—THE OLD RASCAL—BELL-BIRD—STOP AT A SITIO, AND RECEPTION—ORIOLES—A CATTLE SATIO—SWIFT CURRENT—ENTER THE RIO NEGRO—ARRIVE AT BARRA

An unclouded sky was awaiting the sun of the 4th, as we strolled along the river bank, at Serpa, recalling the clustering associations connected with the day, and thinking of the present occupations of friends, at home. It was a magnificent place for fire-works and tar barrels, and that beautiful island opposite, was the very spot for a picnic. We had quite a mind to have a celebration on our own account, for the purpose of demonstrating to the benighted Amazonians how glorious a thing it is to call one's self free and independent; but, alas! our powder was precious, and barrels of tar not to be had for love or money. The sun peeped over the tree tope, flooding in beauty the wild forest, and gilding the waters that rushed and foamed like maddened steeds. The birds were making the air vocal with a hundred different notes, and fishes were constantly bouncing above the water in glee. And was it a fancy, that one red-coated fellow, as he tossed himself up, greeted us with a "viva" to the Independence of America?

Serpa was a pretty place, after all; and our impressions of the night before, had been formed after a long day and a scorching sun. And the people of Serpa were a happy people, and we almost wished that our names were in their parish register. The river teemed with the best of fish, and half an hour's pleasure would supply the wants of a week. Farinha grew almost spontaneously, and fruits quite so. The people bartered with passing boats for whatever else they might require, and lived their lives out like a summer's day, knowing nothing of the care and trouble so busy in the world around them, and happy as language could express. With an income of one hundred dollars, a man would be a nabob in Serpa, as rich as with a hundred thousand elsewhere.

Not far back of the village, is a large lake, the Saracá, and at one of the outlets of this, Mr. McCulloch had, a few years since, made arrangements for a saw-mill. But after several months' labor, when the timbers were all ready to be put together, he was ordered by the authorities at Pará to desist, upon some frivolous pretext. From here, he removed to Barra.

Senhor Manoel had been on the point of leaving for Barra, as we arrived, and he concluded to go with us, putting two of his men upon the galliota. Besides these, we had been unable to find any others. The Colonel and Juiz were also to go in their own canoes, keeping us company. These gentlemen were all going up to Barra to attend a jury, one of the inflictions of civilization in Brazil, as elsewhere. But, although a week's voyaging among the cárapanás is no sport, they did not grumble half so much at the obligation, as many a man, at home, for the loss of his afternoon by similar necessity.

Leaving Serpa, about seven o'clock, we continued on an hour, until we arrived at a spot, whither the Senhors had preceded us, and made ready breakfast. We were to have a picnic after all. Each canoe had brought store of good things, and we circled around a little knoll under the trees, to the enjoyment of a greater variety than we had seen for the last two months.

At this place, we shot an Opossum, of a smaller variety than that of the States. It emitted a very disagreeable odor, and even our Indians expressed their disgust at the idea of eating it. I intended to have preserved it, and laid it in the montaria for that purpose, but soon after, it was missing, some one having thrown it into the stream.

Nearly all day, our course was through a passage, of not more than fifty yards width, between the northern shore and an island. At low water, this channel was entirely dry. In one part of our way, a large flock of Swallow-tailed Hawks, Falco furcatus, a variety found also in the Southern Slates, circled about us in graceful motion, like so many swallows. We brought down one, a fine specimen, greatly to our delight; for although we had frequently seen them before, we never had been able to reach them, on account of their lofty flight.

It was nearly midnight when we reached the sitio of the Delegarde of Serpa, directly opposite the mouth of the river Madeira.

The Colonel had arrived before us, and we found prepared a substantial supper. The Delegarde of Serpa, has not a very lucrative office, and matters about the house looked rather poverty-stricken; but we cared little for that, on our own account, and slinging our hummocks under an open cacao-shed, slept as well as the cárapanás would allow.

The river Madeira is the greatest tributary of the Amazon; having a length of more than two thousand miles. Rising far down among the mountains of Southern Bolivia, it drains a vast extent of country, receiving constant accessions. Its current is not swift, and its waters are comparatively clear. When the Amazon is lowest, in the month of December, the Madeira is at its height; and at that season, very many fallen trees are floated down. Much of the country, about its mouth, is low and uninhabitable; and at certain seasons, the whole region, below the falls, is visited by intermittent fevers. This scourge to man, is a blessing to the turtles, who congregate upon the upper islands, and deposit their eggs, without molestation. The first falls are at the distance of two months' journey from Serpa; and, thereafter, a succession of similar falls and rapids obstructs the navigation for a long distance. Yet canoes of considerable burden, ascend the river passing these falls by aid of the Indians, who are settled about these places in large numbers. By the upper branches of the Madeira, easy communication is had with the head-waters of the La Plata; and in the earlier days of Brazilian settlement, the enterprising colonists had discovered and taken advantage of this connection. To the interior province of Matto Grosso, communication is had by the Tocantins, Tapajos, and Madeira, from Pará. The latter river is preferred, on account of the fewer obstructions, although the distance is greatly increased. Not unfrequently, one of these canoes arrives at the city, loaded with the products of Matto Grosso, among which gold is one of the principal. The Indians, accompanying such craft, are of a very different race from those usually seen; and in strange dresses, wander about the streets, staring at every sight.

There are but few settlements upon the lower waters of the Madeira. The chief of them is Borda, upon the southern bank, two days' voyage from Serpa. The country is rich in woods, cacao, salsa, and gums. A greater obstruction to its settlement than unhealthiness, was the obstinate ferocity of the Indian tribes upon the river banks, especially the Muras and Mundrucús. But both these have yielded in some degree to the effects of civilization, and the latter are now considered one of the most friendly races in the province.

Resuming our journey before daybreak of the 5th, we arrived, about seven o'clock, at the most orderly looking sitio which we had yet seen. There were a number of slaves, and the fields of mandioca and tobacco were as neat as gardens. The houses were well built, and arranged in the form of a quadrangle; and, being upon a lofty bank, commanded a beautiful view of the river, and the remote shore. A grove of orange trees hung loaded with fruit, and we readily obtained permission to fill our lockers. The orange season was just commencing, and thereafter we found them every where in profusion.

Here also we obtained a shell new to us, the Achatina regina.

Three miles above this place, was the village of our Tauçha, and as himself and his party had been absent several months, we observed their demeanor with some curiosity, as we drew near their home. The old man looked sharply, as though he would see if any changes had occurred in his domain; the boys scarcely looked at all, and seemed as apathetic as blocks; but the princess was all smiles, pointing out to her children this and that object, or her recognized friends upon the bank. The village did not present a very distinguished appearance, although upon a singularly fine site; the bank being fifty feet above the water, and fronted by a small island at the distance of a mile. As we touched the shore, a number of women and children were looking on from above, as though we were perfect strangers; only two of the little girls coming down to meet their brothers and cousins. With the same indifference, the boys, as they met their mothers and sisters, scarcely exchanged a salutation. To give them all the credit they deserved, however, their first steps were to the rude chapel, where, before the altar, on bended knees, they thanked our Lady for their safe return. There was one poor boy, the best of the band, who had been sick with jaundice during the whole passage. The others had been perfectly indifferent to him, not caring whether he lived or died; but we had done every thing for his comfort that circumstances would allow, and in return, although he could not speak a word of Portuguese, he had testified his gratitude in a hundred little instances. He lingered about us a long time, as if loath to part; and when, at last, he went upon the hill, where the others were collected together, detailing the wonders of their travels, he slunk away, unnoticed by any; nor did we see the least recognition of him while we remained.

When Lieutenant Mawe descended this river, in 1831, these people had just been gathered out of the woods by an old Padre who had converted them, and taught them something of civilization. Mr. Mawe particularly observes, that they would drink no cashaça, nor exchange fish for that article.

But the old Padre had gone; the houses, far better framed than usual, were almost all in ruins; and there did not seem to be a dozen adults in the place. A large piece of ground had, at one time, been cultivated, but now, the grass and bushes had overgrown the whole; and excepting where a few squash vines had found a home upon the side-hill, not a trace of agriculture remained. With this outward decay, the Padre's instructions had gone likewise, and these Muras were noted as arrant thieves, and lazy vagabonds. The little civilization once acquired, had left behind just enough of its dregs, to make them worse than their brethren of the woods.

We wandered some hours in the vicinity, shell hunting and sporting, with very little success; but the exercise was delightful, for long confinement in the galliota had stiffened our joints, and well nigh put us upon the sick list.

Senhor Manoel Jochin waited until afternoon for the return of some men, who were said to be absent upon a fishing expedition; but, at last, he left, after making the Tauçha promise to forward us with our full complement, when the absentees returned. The Senhor, very kindly, left with us his two men, whom we had employed since leaving Serpa. No sooner was he gone, than the fishermen appeared from the woods, where they had been skulking; and now, the Tauçha, having received payment, refused to do any thing further. There was no help; we could only threaten Doctor Costa's vengeance, and, therefore, prepared to depart as speedily as possible.

The price to be paid this party of six, had been stipulated by Doctor Costa, before their descent. Their wages had been given them in money at Pará, and for the forty-five days, during which they had been in our employ, each received three shirts of factory cotton, three pairs of pantaloons of blue drilling, and two balls of thread. In addition, the Tauçha was to receive at Barra, two whole pieces of drilling; but this, of course, he forfeited by not fulfilling his engagement.

We had still seven men besides the pilot, although we had left eight persons at the village, and were, after all, not so badly off as we might have been.

Bidding adieu to the Muras with uncourteous blessings, we coasted for some hours under the same lofty bank, passing a number of fine sitios. The current was often so swift that the utmost exertions of the men were unable to propel the boat, and they showed great glee at the alacrity with which the Senhors sprang to the paddles for their relief.

During the night, we fancied we heard the far-famed Bell-bird. The note was that of a muffled tea-bell, and several of these ringers were performing, at the same time; some, with one gentle tinkle, others, with a ring of several notes. I asked the pilot, what was "gritando;" he replied, "a toad." I had no idea of having my musician thus calumniated, and remonstrated thereupon; but he cut me short with, "it must be a toad, every thing that sings at night is a toad." From accounts of travelers, we had been expecting, ever since we had entered the Amazon, to have been nightly lulled to sleep by the song of this mysterious bird; and we used, at first, to strain our perceptions to the recognition of something that was bell-like, now starting at the hooting ding-dong of an owl, and now at the slightest twitter of a tree-toad. But it was all in vain; the illusion would not last, and unless, when heart-saddened, his note, which is usually compared to the "pounding of a hammer upon an anvil," comes within the compass of a little bell of silver, we never heard the Bell-bird.

During the whole of the 6th, we were passing through a narrow passage, under a melting sun, and unenlivened by a single bird, or other enticement. An Amazonian sun can be fierce, and upon such days, the birds fly panting into the thickets, and trees and flowers look sorrowfully after them, as though they would gladly follow. The river bank was often high, and occasionally we saw a real rock—no clay fiction.

The cárapanás gave us no rest during the night, and early upon the 7th we were advancing, hoping to arrive at a sitio by breakfast time.

Daybreak found us emerging from our narrow passage, and we saw but a short distance ahead the embarcaçoen, in which most of Bradley's goods had been shipped, and which had left the city, a few days before ourselves. The men pulled lustily to overtake her, for we were out of cashaça, and now should be able to obtain a supply.

It was ten o'clock before we came in sight of the sitio, situated upon a high, projecting bluff. The embarcaçoen was anchored in a little bay upon the upper side. We drew up in a convenient spot below, and walked in procession to the house. The reception chamber in this case, was a raised platform, about two feet high, covered with slats, upon which mats were spread, and over which two hammocks were banging. We found the Senhor and his lady, with the Captain just arrived engaged with their coffee, and the invitation to us was not "entra," but "sobre," that is, "mount." This direction we accurately followed, and squatted ourselves, Turkish fashion, upon the mats. Coffee was presented us, and after our now tasteless galliota preparation, was a luxury.

This house was large enough, and had its proprietor thought fit to limit the circulation of air, by an outer wall or two, or to fetter the grass upon the floor by tiles, would have been one of the finest houses upon the river. But such innovations, probably, never occurred to him. Under the same roof, and within six feet from the platform, was a furnace and anvil, at which, a black Cyclops was officiating, with an earnestness that made our ears a burden, and that puzzled us to comprehend how the good couple could endure their hammocks.

A number of pretty children were playing about, and one of them speedily formed an intimacy with A———. She brought him a cuya of eggs, and seemed happy as a lark, with some trifling present which he made her in return. How often had we wished for some of these pretty toys or books, which children at home value so lightly, but which those upon the Amazon would regard as priceless treasures. Upon leaving, the Senhora sent down half a dozen fowls, and some vegetables for our acceptance.

The proprietor of this establishment was counted one of the wealthiest men upon the river, and we saw numerous slaves, and large fields of tobacco and mandioca. In front of the house, an Indian, and his boy, were weaving a grass hammock, twisting the cord from the raw material as they required it, a few yards at a time.

Soon after starting, we passed the embarcaçoen, obtaining our indispensable. This vessel had large schooner sails, but as wind did not always favor, eight men stood upon her deck, with long sweeps, made by fastening the blades of paddles upon the ends of poles, and pulled her onward. Besides these, two men were in the montaria with a rope, tying and pulling, as before described. In this manner, she advanced nearly as rapidly, or rather, as slowly, as ourselves.

We bad been disappointed in our expectation of obtaining some additional men at this sitio. The riddance of the Tauçha's party was an inconceivable relief; for the men, having no bad example constantly before them, required no urging, but pulled steadily and contentedly from four in the morning until eight at night, frequently cheering their labor by songs. Many of their songs are Portuguese, and the airs are very sweet; but the real Indian is usually unburdened with words, and is little more than a loud, shrill scream, with something of measure; a sort of link between the howl of the performer at the Chinese Museum, and a civilized tone. We never could catch these wild tunes, but they were as natural to every Indian, as his bow and arrow.

We saw a number of beautifully marked Orioles, in orange and black livery, Icterus guttulatus (Lafresnaye), as well as another variety, which we afterwards found to be extremely abundant upon Marajo, Icterus jububa; and the notes of a new variety of Toucan, Pteroglossus aracari, sounded noisily along the shore.

Late at night, we stopped at a cattle sitio. The master was absent, but the slaves had a number of fine Tambaki, and we purchased enough, already roasted, to last us to Barra. Habitual travelers upon the Amazon, make it a point to stop during the night at sitios, whenever possible, thus avoiding the cárapanás, and greatly relieving the tedium of their voyage.

At seven o'clock, upon the 8th, we were in the swiftest current below the Rio Negro. A rocky shore, dry at low water, at this season, formed a rapid, down which the waters rushed with a furious velocity. Two of us went ahead in the montaria; some used the pole; while others, with the sail rope, jumped upon shore and pulled. By these means, after a hard tug, we passed.

We breakfasted in a lovely spot, where the open woods, and the moss-covered rocks, so different from any we had seen before, reminded us strongly of well-loved scenes at home. Here we gathered several species of Ferns, and from a mound of soft red clay, cut out cakes, like soap; for some soil-inquisitive friend.

The remote bank of the Rio Negro now began to rise boldly, exhilarating us all. The water of the Amazon gradually lost its muddy hue, and the black water of the Negro as gradually assumed its proper color; until, at last, intensely dark, but clear and limpid, every ripple sparkling like crystals, it bade us throw back a joyful adios to the majestic old friend we were leaving, and hail with loud vivas the beautiful newly-found.

At its junction with the Negro, the Amazon bends widely to the south; so that, from the northern shore, the former seems the main stream. Directly at the junction, lies a large, triangular island, and Mr. McCulloch informed us, that he himself had found soundings here, at thirty-two fathoms, or one hundred and ninety-two feet. Upon either side, the shore rises abruptly and loftily, and the river is contracted into much narrower limits than above.

We sailed under noble bluffs, passing many fine-looking houses; and the effect of these, with the dark water, the cloudy sky, and the rich green festooning, made that few hours' sail intensely interesting. The current moved sluggishly, and the only signs of life which we met, were in correspondence; a swarthy white, in one end of a montaria, listlessly holding a fish-line, while in the other, sat, curled up, a little boy, in blue shirt and red cap, both pictures of luxurious laziness,

It was eight o'clock in the evening, as we moored to the shore, at Barra. A furious rain was pouring, and thus, we ended our voyage as we had begun it. We had left Pará, expecting to see but thirty days pass upon the Amazon, but the thirty had flown long since, and here we were, upon the eve of the fiftieth.

Yet our time had passed pleasantly, in spite of every inconvenience; and now that the memory of the cárapánas began to fade into indistinctness, and the big flies could no longer trouble us, we could have looked forward to another fifty days towards the Peruvian frontier, without trembling.

The distance from Pará to the Barra of the Rio Negro, in a straight line, is rather more than eight hundred miles, but as we had come, following all the windings of the channel, the distance was more than a thousand.

Early in the morning, a number of gentlemen visited us at the galliota, some to inquire of the market and news below, others to make offers of friendly service. Of these latter was Senhor Henriquez Antonio, an Italian by birth, and the most prominent trader upon these upper rivers. He immediately offered us a vacant house next his own, and in a brief time, we were fairly installed in our new quarters. The building was of one story, containing several rooms, most of which were ceiled by roof tiles, and floored by sand. Bradley took possession of the large parlor for his goods, and he and Mr. Williams were domiciled in one of the little twelve-by-twelve sanctums, and A—— and I in the other.

Chapter XV

RIO NEGRA AT BARRA—THE TOWN—OLD FORT—SR. HENRIQUEZ AND FAMILY—MANNER OF LIVING— VENEZUELANS—PIASSÁBA ROPE—GRASS HAMMOCKS— FEATHER WORK—DESCENT OF THE NEGRO—GALLOS DE SERRS—CHILI HATS—WOODS IN THE VICINITY— TROGONS—CHATTERERS—CURASSOWS—GUANS— PARROTS AND TUUCANS—HUMMING BIRDS—TIGER CATS—SQUIRRELS—A TIGER STORY—THE CASUÉRIS—A YANKEE SAW-MILL—MODE OF OBTAINING LOGS—A PICNIC—CROSS THE RIVER TO A CAMPO—CATTLE AND HORSES—A SELECT BALL

The Rio Negro, at Barra, is about four miles in width, at high water, but much less during the dry season, when the flood has fallen thirty feet. The channel deepens, at once, from the shore, forming a safe and convenient anchorage. The shore, in some parts, is bold, rising in almost perpendicular bluffs; in others, gently sloping to the water's edge. Upon land thus irregular, the town is built, numbering rather more than three thousand inhabitants, a large proportion of which are Indians. The houses are generally of one story, but occasionally of two and three, and resemble, in form and structure, those of the better towns below.

There was something very attractive in the appearance of tile Barra. The broad, lake-like river in front, smooth as a mirror; the little bay, protected by two out-jutting points; the narrow inlet, that circled around the upper part of the town, and beyond which sloped a lofty hill, green with the freshness of perpetual spring; the finely rolling land upon which the town itself stood; and back of all, and overtopping all, the flat table, where, at one glance, we could take in a combination of beauties, far superior to any thing we had yet seen upon the Amazon. Here the secluded inhabitants live, scarcely knowing of the rest of the world, and as oblivious of outward vanities as our Dutch ancestors, who, in by-gone centuries, vegetated upon the banks of the Hudson. Here is no rumbling of carts, or trampling of horses. Serenity, as of a Sabbath morning, reigns perpetual; broken only by the rub-a-dub of the evening patrol, or by the sweet, wild strains from some distant cottage, where the Indian girls are dancing to the music of their own voices.

Directly upon the river bank, and frowning over the waters, once stood a fort, known as San José. The Portuguese word for fort, is barra, and this name was applied to the town which sprung up in the vicinity. Therefore it is, that the town is usually spoken of as the Barra de Rio Negro. Whether peace has been unfavorable, or the fortunes of war adverse, we were not informed; but there stands the ruin, with scarcely wall enough left to call it a ruin, white with lichens, and protecting naught but an area of grass. Upon the top of the ancient flag-staff, is perched a Buzzard, who never seems to move, the livelong day, but to turn his wings to the sun-light, or to nod sympathetically to a party of his brethren, who, upon upright poles and crossbeams, that indicate still further ruin, sit drooping in the "luxury of woe."

Near by, an antique church shoots up to the loftiness of some thirty feet, and at its side, is a quaint adjunct of a tower, square, and short, and thick, from whose top sounds the church-going bell. Beyond this, is a square or Largo, facing which are the Barracks and the Room of the Assembly, for Barra is the chief town of the district of the Rio Negro.

Upon this Largo, stood also the house of Senhor Henriquez, in which we were half domiciled, for being all bachelors, and weary of bachelor cooking, we accepted with pleasure the invitation of Sr. H. to his table. His house was always open to passing strangers, and others beside ourselves were constantly there, enjoying his hospitality. Both the Senhor and his lady, showed us every attention, and seemed particularly anxious that we should see all that was interesting or curious in the vicinity, while they constantly kept some Indian in the woods for our benefit. The Senhora was an exceedingly pretty woman, about twenty-two, and delighted us by her frank intercourse with strangers; always sitting with them at the table, and conversing as a lady would do at home. This would not be noticeable, except in Brazil, and, perhaps, not universally there; but we had ever found the ladies shy and reserved, and, although often at the table of married men, the lady of the house had never before sat down with us. The Senhora surprised and gratified us, also, by her knowledge of the United Slates, which she had obtained from occasional travelers. She had three little girls, Paulina, Pepita, and Lina, with a little boy of four years, Juan. All these children had light hair and fair complexions, and the blue-eyed baby, Lina, especially, was as beautifully fair, as though her home had been under northern skies. Juan was a brave little fellow, and was a frequent visitor of ours, delighting to be with a Gentio Indian, who was employed in our back yard. This Indian had been out of the woods but a few weeks, and could not speak Portuguese, but Juan could talk with him in the Lingoa Geral, as though it had been his native tongue.

Each of the children had an attendant; the girls, pretty little Indians of nine or ten years, and Juan, a boy, of about the same age. It was the business of these attendants, to obey implicitly the orders of their little mistresses and master, and never to leave them. Juan and his boy spent much of their time in the river, taking as naturally to the water as young ducks.

At six in the morning, coffee was brought into our room, and the day was considered as fairly commenced. We then took our guns, and found amusement in the woods until nearly eleven, which was the hour for breakfast. At this meal we never had coffee or tea; and rarely any vegetable excepting rice. But rich soups and dishes of turtle, meat, fish, and peixe boi, in several forms of preparation, loaded the table. The Brazilian method of cooking becomes very agreeable, when one has conquered his repugnance to a slight flavor of garlic, and the turtle oil, used in every dish. The dessert consisted of oranges, pacovas and preserves. Puddings, unless of tapioca, are seldom seen, and pastry never, out of the city. Water was brought, if we asked for it, but the usual drink was a light Lisbon wine. The first movement upon taking our places at the table, was, for each to make a pile of salt and peppers upon his plate, which, mashed, and liquefied by a little caldo or gravy, was in a condition to receive the meat. A bowl of caldo, in the centre, filled with farinha, whence every one could help himself with his own spoon, was always present.

The remainder of the day we spent in preserving our birds, or if convenient, in again visiting the forest. The dinner hour was between six and seven, and that meal was substantially the same as breakfast.

We found at the house, upon our arrival, two gentlemen who had lately came from Venezuela, forty days' distance up the Rio Negro. One of them was a young German, William Berchenbrinck, who had come down merely as passenger, and who had been in the employment of a Spanish naturalist. The other was a regular trader, Senhor Antonio Dias, from San Carlos, and he had brought down a cargo of rope, made from the fibers of the Piassába palm, and a quantity of grass hammocks. The piassába rope is in great demand throughout the province, and is remarkable for its strength and elasticity, which qualities render it admirable for cables. The only objection to it is its roughness, for the palm fibers are, unavoidably, of large size.

The hammocks were, in general, of cheap manufacture, valued at half a milree each. The grass of which they were made is yellow in color, and of a strength and durability superior to Manila hemp. It grows in very great abundance throughout the country of the Rio Negro, and could be supplied to an unlimited extent. Senhor Antonio was a genius in his way, and some of his hammocks were exquisitely ornamented, by himself, with feather work. One, in particular, was composed of cord, twisted by hand, scarcely larger than linen thread; and in its manufacture, a family of four persons had been employed more than a year. Its borders, at the sides, were one foot in width, and completely covered with embroidery in the most gaudy feathers. Upon one side were the arms of Brazil, upon the other, those of Portugal, and the remaining space was occupied by flowers, and devices ingenious as ever seen in needle-work. The feathers were attached to the frame of the borders by a resinous gum. Such hammocks are rather for ornament than use, and they are sought with avidity at Rio Janeiro, by the curiosity collectors of foreign courts. This is one was valued at thirty silver dollars, which, in the country of the Rio Negro, is equal to one hundred, in other parts of the empire.

Sr. Antonio was something of a wag as well as a genius; and as the blacks came to him, at sunset, for the customary blessing, making the sign of the cross upon their foreheads, his usual benediction was, "God make you white."

Berchenbrinck could speak English fluently, and was a very agreeable companion to us, besides being enabled, from his own experience, to contribute much to our information regarding the natural curiosities of the country. He had crossed from the Orinoco to the Rio Negro, by the Casiquiari, and in coming down with Sr. Antonio, had been well nigh drowned in descending one of the many rapids that obstruct this latter river. Their cargo had been sent round by land, but through some carelessness, the vessel had been overturned, and both our friends precipitated into the whirling flood, whence they were, some time after, drawn out, almost insensible, by their crew, who from the shore had watched the catastrophe. Mr. B. informed us, that in the highlands between the two rivers, the Gallo de Serra, or Cock of the Rock, was abundant, and frequently seen domesticated. This bird is the size of a large dove, and wholly of a deep orange color. Upon its head, is a vertical crest of the same. The Indians shoot the Cocks of the Rock with poisoned arrows, and stripping off the skins, sell them to travelers, or traders, who purchase them for feather work. We obtained a number of them at Barra, and had we arrived a short time sooner, could have seen a living specimen, which was in the garden of Sr. Henriquez.

The Indians, who accompanied Sr. Antonio, were of a different race from any we had seen, and looked very oddly, from the manner in which they suffered their hair to grow; shaving it close, except just above the forehead, from which, long locks hung about their cheeks.

One day, an old Spaniard arrived, with a cargo of Chili hats. He was from Grenada, and had come down the River Napo, and the Solemoen. Beside his hats, which he was intending to take to the United States, he brought a quantity of pictures, or rather, caricatures of saints, as small change for his river expenses. Chili hats are a great article of trade at Barra. They are made of small strips of a species of palm, twisted more or less finely. This palm was growing in the garden of Sr. Henriquez, and he gave us a bundle of the raw material. The leaf was of the palmetto form, and looked much like the leaf of which Chinese fans are made. The value of the hats varies greatly, some being worth, even at the Barra, from fifteen to twenty dollars. But the average price is from two to three dollars. We saw one of remarkable fineness, which was cent to Doctor Costa in a letter.

The old Spaniard told us that much of the country upon the Napo was still wild, and that, in repeated instances, the Indians there brought him beautiful birds for sale, which they had shot with poisoned arrows. Two hundred years ago, Acufia described the Tucuna tribe as remarkable for their similar habit

The woods in the vicinity of Barra were a delightful resort to us, and more attractive than we had seen upon the Amazon. The land was not one dead level, swampy, or intersected by impassable igaripés; but there were gentle hills, and tiny brooks of clearest water, and here, when weary of rambling, we could recline ourselves in the delicious shade, unmolested by cárapanás, or the scarcely less vexatious wood-flies. The ground was often covered by evergreens of different varieties, and exquisite forms, and many species of ferns were growing in the valleys. There were no sepaws, or other climbing obstructions to our free passage, but a thousand lesser vines draped the low tree tops with myriads of flowers, new and attractive. Every where were paths, some made by the inhabitants, in their frequent rambles, others, by wild animals that come to the water; and along these, we could pass quietly, to the feeding trees of beautiful birds.

Here were wont to haunt many varieties of Trogons, unknown to us; and, at any hour, their plaintive tones could be heard from the lofty limb, upon which they sat concealed.

Cuckoos of several species, their plumage glancing red in the light, flitted noiselessly through the branches, busied in searching for the worms, which were their favorite food.

Purple Jays, Garrulus Cayanus, in large flocks, like their blue cousins of North America, would be alighted on some fruit tree, chattering and gesticulating; but shy, ready to start at the breaking of a twig.

Motmots, and Chatterers, were abundant as at Pará; the latter, in greater variety, and still, most gaudy of all.

Goatsuckers, in plumage more exquisitely blended, than any of the species we had ever seen, would start from some shade where they had been dozing the day hours, and flying a little distance, were an easy prey.

Manikins were in great variety, and in every bush; Tanagers whistled, and Warblers faintly lisped their notes in the trees.

Flycatchers, in endless variety, were moving nimbly over the branches, or sallying out from their sentry stations, upon their passing prey.

Pigeons, some of varieties common at Pará; others, new to us, were cooing in the thicket, or flying affrighted off.

Tinami, of all sizes, were feeding along the path, or sporting in parties of half a dozen, among the dry leaves.

Curassows moved on with stately step, like our Wild Turkey, picking here and there some delicate morsel, and uttering a loud, peeping note; or ran, with outstretched neck, and rapid strides, as they detected approaching danger.

Guans were stripping the fruits from the low trees, in parties of two and three; and constantly repeating a loud, harsh note, that proved their betrayal.

Of all these birds, the most beautiful, after the Chatterers, were the Trogons. There were half a dozen varieties, differing in size, from the T. viridis, a small species, whose body was scarcely larger than many of our Sparrows, to the Curuqua grande, Calurus auriceps (Gould), twice the size of a Jay. All have long, spreading tails; and their dense plumage makes them appear of greater size than the reality. They are solitary birds, and, early in the morning, or late in the afternoon, may be observed sitting, singly or in pairs; some species, upon the tallest trees, and others, but a few feet above the ground, with tails outspread and drooping, watching for passing insects. Their appetites appeased, they spend the remainder of the day in the shade, uttering, at intervals, a mournful note, well imitated by their common name, Curuqua. This would serve to betray them to the hunter; but they are great ventriloquists, and it is often impossible to discover them, although they are directly above one's head. The species vary in coloring, as in size; but the backs of all are of a lustrous green, or blue, and bellies of red, or pink, or yellow. The Curuqua grande is occasionally seen at Barra; but frequenting the tallest forest, it is exceedingly difficult to be obtained. We offered a high price for a specimen, and employed half the garrison for this single bird, without success. They reported, that they, every day, saw them, and frequently shot at them; but that they never would come down. We were fortunate in obtaining a skin of this bird, preserved by an Indian. The other species were the Red-bellied, T. çurucui; Cinerons, T. strigilatus; T. melanopterus; and one other species, much resembling the last, except that the outer tail-feathers, instead of being merely tipped with white, as in the Melanopterus, were crossed by numerous while bars.

Their feathers were so loose, that, in falling, when shot, they, almost invariably, lost many; and this, together with the tenderness of their skins, made them the most difficult of birds to preserve.

Of the Chatterers, besides the Cardinal, and other Pará varieties, which were beginning to be abundant, were the Pompadour, Ampelis pompadora, whose wings were white, and body of a lustrous carmine; and another variety, the Silky, A. Maynana, whose body was of a sky-blue. At this season, all these birds were in perfect plumage; and seemed to be just returning from their migration, perhaps towards Pará, as they were there during the month of May.

Of Curassows, or Mutuns, we never shot but one variety, the Crested, of which we had found the nests, near Serpa. But other species were common about the forests, and these, with others still, brought from the upper country, were frequently seen domesticated. They are all familiar birds, and readily allow themselves to be caressed. At night, they often come into the house to roost, seeming to like the company of the parrots and other birds. They might easily be bred, when thus domesticated, but the facility with which their nests are found, renders this no object at Barra, They feed upon seeds and fruits, and are considered superior, for the table, to any game of the country. For one patac, or sixteen cents, each, we purchased a pair of the Razor-billed Curassow, Ourax mitu, one of which we succeeded in bringing safely home; a pair of (judging from recollection) the Red-knobbed Curassow, Crax Yarrellii; and a male of the Red Curassow, (Crax rubra), said to have been brought from Peru. This variety was called Urumutun. The second species is the most common, and is found throughout the country, towards Pará, The Parraqua Guan, Phasianus parraqua, was common, but not domesticated. It resembled the Mutuns in its habits, but in form, had a larger neck and tail, in proportion. A specimen which we shot, exhibited a very curious formation of the wind-pipe, that organ passing beneath the skin, upon the outside of the body, to the extremity of the breast-bone, where it was attached by a ligament. Then re-curving, it passed back, and entered the body as in other birds. Probably, the loud trumpet note of this bird is owing to this formation.

Of Parrots and Toucans there were many new varieties, besides some of those common at Pará. One species of Parroquet was scarcely larger than a canary bird.

Of Hawks there were many varieties, not known at Pará, and a large long-eared Owl, the first owl we had met, was brought in by our hunters.

Humming birds were abundant as elsewhere, but mostly of species observed at Pará. The Amethystine, T. amethystinus; and the Black-breasted, T. gramineus, were all the new varieties that we obtained.

Our hunters were mostly soldiers of the garrison, and for their labor we paid them ten cents per diem, and found them in powder and shot. When, towards night, they made their appearance with the fruits of their excursions, our table was richly loaded, and a long evening's work spread before us.

Sometimes, they would bring in animals, and upon one occasion, we received a pair of small Tiger Cats, called Máracajás.

Some varieties of Squirrels were also brought in, but as we had no leisure to attend to animals, we gave no orders for procuring them. The same animals found in other parts of the province, were common in the vicinity, and we could learn of nothing new, excepting Monkeys, who vary in species with every degree of latitude or longitude.

Mr. McCulloch gave us the teeth of a Jaguar, which he had shot at his mill; and we heart of a singular meeting between one of these animals and an Indian, upon the road towards the mill. The Jaguar was standing in the road, as the Indian came out of the bushes, not ten paces distant, and was looking, doubtless, somewhat fiercely, as he waited the unknown comer. The Indian was puzzled an instant, but summoning his presence of mind, he took off his broad brimmed hat, and made a low bow, with "Muito bem dias, meu Senhor," or "a very good morning, sir." Such profound respect was not lost upon the Jaguar, who turned slowly, and marched down the road, with proper dignity.

Several times, during the latter part of our stay, when our names had acquired some celebrity, birds, and other curiosities were brought in for sale; and, upon one day in particular, such a zeal for vintence actuated all the little blackies and Indians, that our big bellied bottles speedily became crowded to repletion, with beetles, and lizards, and snakes, *et id omue genus*.

Three miles back of Barra, is the Casuéris, a water fall, of which Mr. McCulloch has taken advantage for his mill. The water falls over a ledge of yellowish red sand rock, and, during the dry season, has a descent of twelve feet. But during the wet season, the waters of the Rio Negro set back to such an extent, that a fall is scarcely perceptible. These changes have their conveniences, for as, when the water is low, the wheel can be constantly turning, so, when it is high, the supply of logs can be floated directly to the mill. The greater part of the logs used, are of cedar, rafted up from the Solimoen. Coming from the head waters of the various streams, they are precipitated over cataracts, and rolled and crushed together, until their limbs are entirely broken off, and their roots require but little trimming. Logs of other woods are cut upon the banks of the Rio Negro, and from low land, during the dry season. When the waters rise, these logs are floated out, bound together, and rafted down. We saw a variety of beautiful woods; some of the most valuable of which for cabinet purposes, were the Saboyerana, reddish, mottled with black, and varieties of Satin-wood. These are scarcely known down the river, but through Mr. McCulloch's enterprise, they are in a fair way to be made common. The mill was a perfect Yankee mill, differing, in no respect, excepting in the materials of its frame; woods beautiful as mahogany not being so accessible as hemlock, in the United Slates.

Heretofore, all the boards used in the province of Pará, have been hewn in the forest, by the Indians, who are remarkably expert at this kind of work, using a small adze, like a cooper's hammer, and making the boards as smooth as with a plane. One log will make but two boards, and the labor of reducing to the requisite thinness is so tedious, that very few builders can afford to use wood for the flooring of their houses. But these people are so proverbially slow in adopting innovations, that some years must elapse, before this expensive system is changed.

The Casuéris being a delightful spot, shaded by densely leaved trees, is the usual resort for Sunday picnic parties, which meet there for the fresh, cool air, and the luxurious bath. The Senhora Henriquez made a little party of the kind for our entertainment, which passed off delightfully, and much as such a party would have done at home. It was something novel, to meet such an evidence of refinement so far out of the world, where we had expected to find nothing but wildness. But there was one feature that distinguished it from any pleasure party I ever participated in, amid civilization and refinement, and that was, the bathing, at the finale. In this, there was little fastidiousness although perfect decorum. While the gentlemen were in the water, the ladies, upon the bank, were applauding, criticizing, and comparing styles, for there were almost as many nations of us, as individuals; and when, in their turns, they darted through the water, or dove, like streaks of light, to the very bottom, they were in nowise distressed that we scrupled not at the same privilege. They were all practiced and graceful swimmers, but the Senhora particularly, as she rose, with her long hair, long enough to sweep the ground when walking, enshrouding her in its silken folds, might have been taken for the living, new-world Venus.

For bathing purposes, we never saw water that could compare with the Rio Negro. One came from its sparkling bosom, with an exhilaration, as if it had been the water of a mineral spring. In it, the whole town, men, women, and children, performed daily ablutions, cleanliness being a part of the Brazilian religion. The women were usually in before sunrise, and we never saw, as some have asserted is the case, both sexes promiscuously in the water.

We crossed the river, one day, in a montaria, with three Indians, to visit a large campo. Our last mile was through woods, the low shrubbery of which was entirely overflowed, and as far down as we could see, were trees in full leaf, looking like a bed of green. Many creeping plants, bearing a profusion of flowers, overhung our heads, and of the finest, a Dendrobium, with its clusters of pink and purple, we obtained a specimen, which we were fortunate enough to bring safely to the United States. In this retreat, we observed a great number of Trogons and Doves, as though the water-side was their favorite resort. The trunks of the trees were all marked by the waters of the last year, full five feet above their ordinary rise. That unprecedented flood poured over the low lands, and caused great devastation.

The campo was some miles in length, covered with grass and low shrubs. The late dryness had deprived the grass of all its green, and the whole resembled more a desert than a meadow. There were a number of lean cattle and horses wandering about, looking for food, with microscopic eyes.

Cattle are rare at Barra, and we saw no milk during our stay. There was said to be one horse, but he was altogether beyond our ken; and the honors of his genus were done by three asses, who were outrageous vagabonds, and unfair proxies.

A ball was got up, for our especial advantage and honor, one evening. Six ladies, some well dressed, some so-so; some tolerably white and some as tolerably dark, composed the lively part, and about a dozen gentlemen, an essential part, of the gathering. One gentleman volunteered to the guitar, another to the violin; one and another sent in refreshments, and an old lady took in charge the coffee. The ladies were very agreeable, differing mightily from the ladies at Pará dancing parties, who do not go to talk. The dances were waltzes, cotillions, and fandangos, and some of the ladies danced with extreme grace. Those who were deficient in grace, made up in good will, and until a late hour, all went on merrily and delightfully.

Chapter XVI

*A NEW RIVER—RIO BRANCO—TURTLE WOOD—
UNEXPLORED REGION—TRADITIONS—PEIXE BOI OR
COW FISH—TURTLES—INFLUENCES AT BARRA—
INDIANS—FOREIGNERS—INDIAN ARTICLES—POISON
USED UPON ARROWS—TRAFFIC—BALSAM COPAIVI—
SALSA—QUINIA—VANILLA—TONGA BEAMS—
INDIGO—GUARANÁ—PIXIRI OR NUTMEG—SERINGA—
WILD COTTON—ROCK SALT—THE AMAZON ABOVE THE
RIO NEGRO—THE RIO NEGRO*

While we were at Barra, Senhor Gabriel, one of the dignitaries of the place, and a very agreeable gentleman, returned from an exploring expedition, up one of the smaller rivers, which flow into the Rio Negro, between Barra and the Branco. Nothing had previously been known of the region lying adjacent to this stream, for vague traditions of hostile Indiana had deterred even the adventurous frontiers-men, from attempting its exploration. The Senhor described it as a beautiful, rolling country, in many parts high, and covered by forests of magnificent growth. It was uninfested by cárapanás, and never visited by fevers; nor were there troublesome Indians to molest settlers.

The Senhor gave us the skin of a large black monkey, which he had killed during this excursion, and the nest and eggs of a White-collared Hummer, the Trochilus melivorus. The nest was composed of the light down growing upon the exterior of a small berry, and surpassed any thing we had seen in bird architecture. The eggs were tiny things, white, with a few spots of red.

The Rio Branco is another interesting stream, which sends if wealth to Barra. Its head waters are in the highlands, towards Guiana, and it flows through one of the loveliest and most desirable regions of tropical America. There are many settlements upon its banks, and an extensive traffic is carried on in cattle and produce. Far up among the mountains, at the head of this river, is found the Márapaníma, or Turtle wood, specimens of which may sometimes be seen made into canes. This is the heart of a tree, and is never more than a few inches in diameter. The only person who deals in it upon the Branco, is a Friar, who obtains it from some Indian tribe, in the course of his mission, and, a few sticks at a time, he sends it to Pará, where it is in great demand for canes, and other light articles. In the same district, are said to be valuable minerals, and we obtained of a canoe which had just come down, a piece of red jasper, susceptible of a fine polish, which was used as a flint. We saw, also, some large and beautiful crystals, from the same highlands.

The whole region, north of the Amazon, is watered by numberless rivers, very many of which are still unexplored. It is a sort of bugbear country, where cannibal Indians and ferocious animals abound to the destruction of travelers. This portion of Brazil has always been Fancy's peculiar domain, and, even now, all kinds of little El Dorados lie scattered far, far through the forest, where the gold and the diamonds are guarded by thrice horrible Cerberi. Upon the river banks are Indians, watching the unwary stranger, with bended bow, and poisoned arrow upon the string. Some tribes, most provident, keep large pens skin to sheepfolds, where the late enthusiastic traveler awaits his doom, as in the cave of Polyphemus. As if these obstructions were not enough, huge, nondescript animals add their terrors, and the tormented sufferer, makes costly vows that if he ever escapes, he will not again venture into such an infernal country, even were the ground plated with gold, and the dew drops priceless diamonds. Some naturalist Frenchman, or unbelieving German, long before the memory of the present generation, ventured up some inviting stream, and you hear of his undoubted fate, as though your informant had seen the catastrophe. In instances related to us, no one seemed to allow, that one might die, in the course of nature, while upon an expiring expedition, or that he might have had the good fortune to have succeeded, and to have penetrated to the other side.

We heard, one day, that a Peixe boi, or Cow-fish; had just arrived in a montaria, and was lying upon the beach. Hurrying down, we were just in time to see the animal before he was cut up. He was about ten feet in length, and as he laid upon his back, between two and three feet in height; presenting a conformation of body, much like that of a "fine old English gentleman," whose two legs were developed into a broad, flat tail. His back was covered sparsely with hairs, and his large muzzle was armed with short, stiff bristles. His smooth belly was bluish-black in color, and much scarred by the bite of some inimical fish. There was nothing corresponding to legs, but a pair of flappers, as of a turtle, answered his purposes of locomotion. Both eyes and ears were very small, but the nostrils were each an inch in diameter. The skin was one-fourth of an inch in thickness, and covered a deep coating of blubber, the extracted oil of which is used as butter in cooking. Under the blubber was the meat, something between beef and pork, in taste. These curious animals are in great numbers upon the Solemoen, and are to the people, what Periecu is below, being, like that fish, cut into slabs and salted. This form is, however, very offensive to a stranger, and no Indian will eat dried peixe boi, if he can get any thing else. These animals do not venture upon land, but subsist upon the grass that lines the shores. When thus feeding, they are lanced by the Indians, who know their places of resort, and watch their appearance. Although from their bulk, several men might be puzzled to lift a cow-fish from the water, when dead, yet one Indian will stow the largest in his montaria, without assistance. The boat is sunk under the body, and rising, the difficult feat is accomplished.

Not unfrequently, a peixe boi is taken eighteen feet in length. Their thick skins formerly served the Indians for shields, and their jaw-bones as hammers.

We would gladly have bought this entire animal, for the purpose of preserving his skeleton and skin. But as meat was in request that day, we were obliged to be content with the head, which we bore off in triumph; and cleansed of its muscle. This skull is now in the collection of Dr. Morton, and we learn from him that the Peixe boi of the Amazon is a distinct species from the Manatus, sometimes seen in the districts adjacent to the Gulf of Mexico.

Sometimes young cow-fishes are brought to Pará, and we had there previously seen one in a cistern, in the palace garden. It was fed on grass, and was very tame, seeming delighted to be handled. Captain Appleton, who has taken greater interest in the wonders of this province, than almost any person who ever visited Pará, has twice succeeded in bringing young cow-fishes to New York, but they died soon after leaving his care.

The Turtles are a still greater blessing to the dwellers upon the upper rivers. In the early part of the dry season, these animals ascend the Amazon, probably from the sea, and assemble upon the sandy islands and beaches, left dry by the retiring waters, in the Japúra and other tributaries. They deposit their eggs in the sand, and at this season, all the people, for hundreds of miles round about, resort to the river banks as regularly as to a fair. The eggs are collected into montarias, or other proper receptacles, and broken. The oil, floating upon the surface, is skimmed off, with the valves of the large shells found in the river, and is poured into pots, each holding about six gallons. It is computed that a turtle lays one hundred and fifty eggs in a season. Twelve thousand eggs make one pot of oil, and fix thousand pots are annually sent from the most noted localities. Consequently, seventy-two millions of eggs are destroyed, which require four hundred and eighty thousand turtles to produce them. And yet but a small proportion of the whole number of eggs are broken. When fifty days have expired, the young cover the ground, and march in millions to the water, where swarms of enemies, more destructive than man, await their coming. Every branch of the Amazon is resorted to, more or less, in the same manner.; and the whole number of turtles is beyond all conjecture. As before remarked, those upon the Madeira are little molested, on account of the unhealthiness of the locality in which they breed. They are said to be of a different and smaller variety, from those upon the Amazon. We received a different variety still from the Branco, and there may be many more yet undistinguished. The turtles are turned upon their backs, when found upon the shore, picked up at leisure, and carried to different places upon the river. Frequently, they are kept the year round, in pens properly constructed, and one such, that we saw at Villa Nova, contained nearly one hundred. During the summer months, they constitute a great proportion of the food of the people; but when we consider their vast numbers, a long period must elapse before they sensibly diminish. Their average weight, when taken, is from fifty to seventy-five pounds, but many are much larger. Where they go, after the breeding season, no one knows, for they are never observed descending the river; but, from below Pará, more or less are seen ascending, every season. They are mostly caught, at this time, in the lakes of clear water, which so plentifully skirt either shore, and generally are taken with lances, or small harpoons, as they are sleeping on the surface. But the Muras have a way of capturing them, peculiar to themselves; shooting them with arrows, from a little distance, the arrow being so elevated, that in falling, it strikes, and penetrates the shell. In this, even long practice can scarcely make perfect; and fifty arrows may be shot at the unconscious sleeper before he is secured.

There are several other small varieties of Turtles or Terrapins, somewhat esteemed as food, but in no request. Some of them are of curious form, and one in particular, found about Pará, instead of drawing in his head and neck, as do most others of his family, finds sufficient security by laying them round upon his fore claw, under the projecting roof of shell.

The land turtles, Jabatis, attain a size of from twenty to thirty pounds. They are delicious food, far superior, in our estimation, to their brethren of the water. Lieutenant Mawe somewhere remarks to this effect, that, in a country where the people are cannibals, and eat monkeys, they might enjoy land turtles. But the Lieutenant suffered his prejudices to run away with his judgment, in a strange way for a sailor.

We saw at Senhor Bentos' in Villa Nova, turtles of this species, which he had in the yard as pets, and who seemed very well domesticated, eating pacovas, or any sweat fruit. Some of these, the Senhor had kept for seven years, and they bore no proportion in size to others seen. From this, we inferred the great number of years that they must require before they arrive at maturity.

Owing to its remote frontier position, Barra is under different influences from other Brazilian towns, and these are observable every where. The language spoken is a patois of Portuguese and Spanish, with no very slight mixture of the Lingoa Geral. This latter language must be spoken, as matter of necessity. The currency, too, is in good part of silver, as Spanish dollars, the Brazilian paper being but in scanty supply.

The Indian population is vastly more numerous than below, and from the absence of the causes that elsewhere have driven the Indians to the woods, the two races live together amicably, and will, to all appearance, in a few generations, be entirely amalgamated. Labor, of course, is very cheap. Senhor Henriquez had one hundred Spanish Indians in his employ, to whom he paid twelve and one-half cents each per diem. These were hired of the authorities beyond the frontiers, and they were protected, by contract, from being sent below Barra. They were of a darker color, and less finely featured than most Brazilian Indians, whom we had seen. Part of them were employed in building houses, several of which were in progress of erection; and part in a tilaria, within the town. When Lieutenant Smythe descended the Amazon, rather more than ten years since, both houses and tilaria were in a sad state; and the town was nearly stripped of inhabitants, on account of recent political difficulties. But better times have come, and a general prosperity in rapidly removing the appearances of decay.

There were a great many pleasant people, whose acquaintance we made, and who showed us such attentions as strangers love to receive. There are always, in such towns, a few strange wanderers from other countries, who have chanced along, no one knows how. Such an one was a German we found there, Senhor Frederics. He had formerly belonged to a German regiment, which was stationed at Pará, and had been lucky enough to escape the fate of most of his comrades, who had been killed during the revolution. He had found his way to the Barra, had married a pleasant lady of the place, and now practiced his trade as a blacksmith. He was a man of tremendous limb, and with a soul in proportion, and we were always glad to see him at our house. Another German was a carpenter; and an odd genius, from the north of Europe, but who had been a sailor in an English vessel, and had picked up a collection of English phrases, officiated as sail-maker to the public.

Through the kindness of Senhor Henriquez, we obtained a great variety of Indian articles. The bows and lances are of some dark wood, and handsomely formed and finished. The former are about seven feet in length, and deeply grooved upon the outer side. The bow-string is of hammock grass. The lances are ten feet long, ornamented with carvings at the upper extremities, and terminated by tassels of macaw's feathers. The arrows are in light sheaves, six to each, and are formed of cane, the points being of the hardest wood, and poisoned. These are used in war and hunting, and differed from the arrows used in taking fish, in that the points of the latter are of strips of bamboo or bone. Those for wild hogs again, are still different, being terminated by a broad strip of bamboo, fashioned in the shape of a pen. This form inflicts a more effectual wound. In the same way, the javelins are pointed, the stems being of hard wood, and much ornamented with featherwork.

But the most curious, and the most formidable weapon, is the blowing-cane. This is eight or ten feet in length—two inches in diameter at the larger end, and gradually tapering to less than an inch at the other extremity. It is usually formed by two grooved pieces of wood, fastened together by a winding of rattan, and carefully pitched. The bore is less than half an inch in diameter. The arrow for this cane, is a splint of a palm, one foot in length, sharpened, at one end, to a delicate point, and, at the other, wound with the silky tree-cotton, to the size of the tube. The point of this is dipped in poison, and slightly cut around, that, when striking an object, it may break by its own weight, leaving the point in the wound.

With this instrument, an Indian will, by the mere force of his breath, shoot with the precision of a rifle, hitting an object at a distance of several rods. Our Gentio Pedro never used any other weapon; and we saw him, one day, shoot at a Turkey Buzzard, upon a housetop, at a distance of about eight rods. The arrow struck fairly in the breast, the bird flew over the house, and fell dead. Senhor Henriquez assured us that an Indian formerly in his employ, at one time and another, had brought in seven Harpy Eagles, thus shot.

The accounts we received of the composition of this poison were not very explicit, and amounted principally to this; that it was made by the Indians at the head waters of the Rio Branco, from the sap of some unknown tree; that it was used universally by the tribes of Northern Brazil, in killing game, being equally efficacious against small birds and large animals; that the antidotes to its effect were sugar and salt, applied externally and internally. It comes in small earthen pots, each holding about a gill, and is hard and black, resembling pitch. It readily dissolves in water, and is then of a reddish brown color. Taken into the stomach, it produces no ill effects. We brought home several pots of this poison, and by experiments, under the superintendence of Dr. Trudeau, fully satisfied ourselves of its efficacy. The subjects were a sheep, a rabbit, and chickens. The latter, after the introduction of one or two drops of the liquid poison into a slight wound in the breast or neck, were instantly affected, and in from two to three minutes were wholly paralyzed, although more than tea minutes elapsed before they were dead. The rabbit was poisoned in the fore shoulder, and died in the same manner, being seized with spasms, and wholly paralyzed in eight minutes. The effect upon the sheep was more speedy, as the poison was applied to a severed vein of the neck.

As M. Humboldt witnessed the preparation of the poison, and has given a full account of his observations, his recital will here not be out of place. The Indian name is Curaré. It is made from the juice of the bark and the contiguous wood of a creeping plant, called the Mavacuré, which is found upon the highlands of Guiana. The wood is scraped and the filaments mashed. The yellowish mass resulting is placed in a funnel of palm leaves; cold water is poured upon it, and the poisonous liquid filters drop by drop. It is now evaporated in a vessel of clay. There is nothing noxious in its vapor, nor until concentrated, is the liquid considered as poisonous. In order to render it of sufficient consistence to be applied to the arrows, a concentrated, glutinous infusion of another plant, called Kiracaguero, is mixed with it, being poured in while the curaré is in a state of ebullition. The resulting mixture becomes black, and of a tarry consistence. When dry, it resembles opium, but upon exposure to the air, absorbs moisture. Its taste is not disagreeable, and unless there be a wound upon the lips, it may be swallowed with impunity. There are two varieties, one prepared from the roots, the other from the trunk and branches. The latter is the stronger, and is the kind used upon the Amazon. It will cause the death of large birds in from two to three minutes, of a hog in from ten to twelve. The symptoms in wounded men ate the same as those resulting from serpent bites, being vertigo, attended with nausea, vomitings, and numbness in the parts adjacent to the wound. It is the general belief that salt is an antidote, but upon the Amazon, sugar is preferred.

The Indian stools were curious affairs, legs and all being cut from the solid block. The tops were hollowed to form a convenient seat, and were very prettily stained with some dye.

Beside these things, were various articles woven of cotton, and of extreme beauty; sashes, bags, and an apparatus worn when hunting, being a girdle, to which were suspended little pouches for shot and flints.

The civilized Indians rarely use their ancient weapons, except in taking fish. Cheap German guns are abundant throughout the country, and it is wonderful that accidents do not frequently occur, with their use. Unless a gun recoils smartly, an Indian thinks it is worth nothing to shoot with; and we knew of an instance, where a gun was taken to the smith's, and bored in the breech, to produce this desirable effect.

Senhor Henriquez has establishments upon several of the upper rivers. Coarse German and English dry goods, Lowell shirtings, a few descriptions of hardware, Salem soap, beads, needles, and a few other fancy articles, constitute a trader's stock. In return, are brought down, balsam, gums, wax, drugs, turtle oil, tobacco, fish, and hammocks.

When Sr. H. goes to Ega, a distance of less than four hundred miles, he forwards a vessel thirty days before his own departure, intending to overtake it before its arrival. So tedious is navigation.

The quantity of Balsam Copaiva brought down, is prodigious. There were lying upon the beach, at Barra, two hollowed logs, in which balsam had been floated down from above. One had contained twenty-five hundred, and the other sixteen hundred gallons. They had been filled, and carefully sealed over; and in this way, had arrived without loss, whereas, in jars, the leakage and breakage would have been considerable. At Barra, the balsam is transferred to jars, and shipped to the city. There, much of it is bought up by the Jews, who adulterate it with other gums, and sell it to the exporters. It is then put up in barrels, or in tin, or earthen vessels, according to the market for which it is intended.

The tree grows in the vicinity of Barra, and we were very desirous of obtaining, at least, some leaves; but delay of one day after another, at last made it impossible. The tree is of large size, and is tapped by a deep incision, often to the heart. In this latter case, the yield is greater, but the tree dies. The average yield is from five to ten gallons.

Sarsaparilla, is another great article of production. It is found throughout the province; and when collected and carelessly preserved, is packed in so rascally a manner, as to destroy its own market. We saw some, that was cultivated in a garden; and its large size and increased strength showed, clearly enough, that by proper care, the Salsa of Pará might compete with the best, in any market. It is a favorite remedy in the country; and when fresh, an infusion of it, sweetened with sugar, forms an agreeable drink.

Quinia grows, also, pretty universally. Happily, for intermittent fevers, opportunities rarely occur of testing its qualities. We never encountered but one case of this fever, which we were enabled to relieve by a single dose from our medicine box.

Vanilla, grows every where; and might, by cultivation, be elevated into a valuable product.

Tonga beans are brought to Barra, from the forest.

Indigo, of superior quality, is raised in sufficient quantities for home consumption; and might be, to any extent.

Not far from Barra, is obtained the nut of which Guaraná is made; which article is extensively consumed throughout the greater part of Brazil, in the form of a drink. The plant is said to produce a nut, shaped somewhat like a cherry; and this is roasted, pounded fine, and formed into balls. A teaspoonful, grated into a tumbler or water, forms a pleasant beverage; but when drank to excess, as is generally the case, its narcotic effects greatly injure the system. The grater, used for this and other purposes, is the rough tongue-bone of one of the large river fish.

There is another fruit, called Pixiri, considered as an admirable substitute for nutmeg. It is covered with a slight skin, and when this is removed, falls into two hemispherical pieces. Its flavor is rather more like sassafras, than nutmeg.

Seringa trees abound upon the Amazon, probably, to its head waters. The demand for the gum has not yet been felt at Barra, where it is only used for medicinal purposes, being applied, when fresh, to inflammations. But when it is wanted, enough can be forthcoming to coat the civilized world.

The Sumaumeira tree, which yields a long-stapled, silky, white cotton, grows upon the banks of the Rio Negro, in great abundance; and could probably be made of service, were it once known to the cotton-weaving communities. It is excessively light, flying like down; but the Indians make beautiful fabrics of it.

Another article, which might be made of inestimable value to the country, is Salt. Upon the Huallaca, and perhaps other tributaries, are hills of this mineral, in the rock; and so favorably situated, as to fall, when chipped off, directly upon the rafts of the Indians, who collect it, and bring it as far down as Ega. It sometimes finds its way to Barra, and we were fortunate in obtaining a piece, weighing nearly one hundred pounds. It is of a pinkish color, and is impregnated with some foreign substance, that needs to be removed. Some enterprising Yankee will make his fortune by it yet. All the salt now used, throughout an area of one million square miles, is imported from Lisbon, and at an enormous expense.

Before closing this chapter, a brief mention of the principal towns, and of the larger rivers above the Negro, may not be inappropriate. At a distance of one hundred miles from Barra, enters the river Perus, a mighty stream, flowing from the mountains of Bolivia. We were informed by individuals, who had voyaged upon this river, that its course was more winding than any other; that it was entirely unobstructed by rapids, and, therefore, preferable to the Madeira, as a means of communication with the countries upon the Pacific. Its banks abound in seringa trees; and cacao, of good quality, is brought down by traders.

Three hundred miles above Barra, is the town of Ega, upon the southern side of the Amazon. It stands upon a river of clear water, which is navigable for canoes, to a distance of several hundred miles; but for larger vessels, but a few days' journey. The town contains about one thousand persons. Upon the northern side, comes in the Japúra, through many channels. This river rises in the mountains of New Grenada, and its broad channel is sprinkled with a thousand islands. During the wet season, it is one of the greater branches of the

Amazon, and flows with a furious current; but during the dry season, it is so filled with sandy shoals, that navigation is impossible. Here the turtles frequent, and down the torrent come vast numbers of cedars. The Japúra, is said to have communication with the Negro, by some of its upper branches. It forms the line of boundary between the Spanish and Brazilian territories. Its region is considered unhealthy; and owing to this reputation, and the obstructions to navigation, is little settled by whites.

Opposite one of the mouths of the Japúra, is the little town of Fonteboa, one hundred miles above Ega. The rivers flowing into the Amazon in this vicinity are numerous, and large, but their courses are said to be laid down upon maps, with the greatest inaccuracy.

The most remote town is Tabatinga, on the northern bank, opposite the mouth of the Javari. This town contains but a few hundred inhabitants. Its distance from Pará is from sixteen to eighteen hundred miles, a six months' journey for the river craft. The country between Tabatinga and the Madeira was formerly inhabited by a tribe called Solimoens, and that part of the river between the Negro and Ucayali, is called by their name.

Beyond the Brazilian frontiers, enter many great branches, the Napo, the Marañon, or Tunguragua, and the Ucayali. The latter is considered the main stream, and down its western branch, the Huallaca, Messrs. Smythe and Lowe came in 1834, starting from Lima. They were in search of a navigable communication between the two Oceans, but were unsuccessful. Whether such a stream exists, as, by a few miles portage, would answer this purpose, is problematical. The country has never been thoroughly explored. The depth of the Amazon, for a long distance up the Ucayali, is very great; at every season navigable for steamboats, unobstructed by rapids, snags, or sawyers.

The Negro receives, in its course, about forty tributaries, and from the healthiness of the region through which it flows, has long been a favorite resort of settlers. A greater number of towns are upon its banks, than upon any other branch of the Amazon. At nine days' distance from Barra, is the town of Barcellos, formerly the capital of the District of the Rio Negro. Eight days beyond this, are rapids, and these are found in succession, for a distance of twenty days. At forty days' distance from Barra, is the Casiquiari, the connecting stream with the Orinoco. Its passage is frequently made, and we encountered several persons who had crossed from Angostura.

Chapter XVII

PREPARE TO LEAVE BARRA—DIFFICULTY IN OBTAINING MEN—THE MAIL—KINDNESS OF OUR FRIENDS—RE-ENTER THE AMAZON—ARRIVE AT SERPA—A DESERTION—WORKING ONE'S PASSAGE—DISORDERLY BIRDS—PASS TABOCAL—SNAKE-BIRD—MARAKONG GEESE—BREEDING PLACE OF HERONS—ARRIVE AT VILLA NOVA—THE COMMANDANT—VISIT TO THE LAKE—BOAT BUILDING—MILITARY AUTHORITIES—SCHOOL—KING OF THE VULTURES—PARTING WITH SR. BENTOS—PASS OBIDOS—CARACARA EAGLE—OUR CREW—INDIAN NAME OF THE AMAZON

After twenty days had passed delightfully, we prepared to leave the Barra, upon the 28th of July, in the galliota, which was to return for Doctor Costa, who was probably awaiting us at Pará. Senhor Pinto, the Delegarde, had promised us some Indians, and another official had assured us of others; but it was discovered, when upon the beach, at the last moment, that both had counted upon the same men. These were three of the Villa Nova police, who happened to be up, and with our Gentio, Pedro, and other whom Senhor Henriquez lent us, were all we could muster. They were less than half our complement, and none of them were to go below Villa Nova. We had letters to the commandant of that place, and he was to provide men for our further advance, in consideration of our being the bearers of His Majesty's mail, and of dispatches from Venezuela. This mail proved a great acquisition, and I would advise all travelers upon the Amazon to secure the same charge.

It was three o'clock, in the afternoon, when our friends gathered upon the beach to bid us adieu. From all of them, although our acquaintance had been so very brief, we were sorry to part; but from Senhor Henriquez, to whom we had been under a thousand obligations, and from Mr. Bradley and Mr. Williams, who had so long been our companions, and to whom we were the more closely drawn, from our being strangers together, in a strange land, the last embrace was peculiarly painful. Messrs. McCulloch and Sawtelle had left some days previously, for the upper waters of the Rio Negro. We had said adios to the Senhora Henriguez an hour before, and her husband told us, that after our departure from the house, she had sat down to a quiet little weep on our account.

The kind lady had sent down to the galliota, a store of meat and chickens, sufficient for some days to come, besides a large basket of cakes made of tapioca, and a turtle. To these, she had added half a dozen parrots and parroquets, as companions of our voyage.

Senhor Pinto had had a large basket made, and in it were a pair of the beautiful geese of the country, Chenalopix jubatua (Spix), called Marakonga, and a Yacou Guan, a rare species, from the country above. With these was also a Red and Yellow Macaw, who was unusually tame, and promised to keep the parrots in subjection. Most of our mutune we were obliged to leave behind, for want of room; and a tiny monkey, which we had bought for a lady friend at home, was retained by his rascally master, on the plea that he was in a tree in the yard, and that he could not catch him.

Barra quickly disappeared from view, and before dark, we were floating down the Amazon, at the rate of about four miles an hour. There were but two of us, and we were just enough to fill the cabin comfortably, reserving any spare corners for our collections of one article and another, and for any of the respectably behaved parrots. The geese and their basket, were slung by the side of the cabin; and the macaw was elevated upon a cross in front of the tolda. Below, were several logs of beautiful woods; and a few bags of coffee, which some friend had shipped for Santarem. A few turtles found space to turn themselves, among the rest, and answered well as ballast. The sail was left behind, as we had no further use for it, the wind generally blowing strongly from below.

In the middle of the stream, cárapanás did not molest us, and we slept through the night as quietly, as if at home. There was no danger of encountering snags, or floating logs, and therefore we kept no watch, but let the boat drift down stern foremost.

Early upon the 29th, we passed the mouth of the Madeira, and shortly after, the village of our old Tauçha. A number of people were upon the hill, and seemed beckoning us to stop, but we were not desirous of farther intimacy with his highness or any of his subjects. When upon better terms, the old man had very politely invited us to stop a few days with him, upon our descent, and had promised us great assistance in collecting birds and shells.

Before daybreak, upon the 30th, we were moored off Serpa. Here we had hoped to obtain additional men, but Senhor Manoel Jochin was absent, upon the Madeira, and excepting one petty officer, and a few soldiers, not a man was left in the place. Senhora Jochin commiserated our situation, and offered to enlist a complement of women, but this was too terrible to think of. She sent us some roasted chickens, eggs, and pacovas; and as we had nothing further to detain us, we cast loose from Serpa.

Meanwhile, two of our policemen had taken their montaria and deserted, leaving us with but three men. This number was hardly sufficient to keep the boat in its course, but, fortunately, there was little wind. A—— and I took our turns at the helm, and we soon discovered, that however romantic the working one's passage down the Amazon might seem, at a distance, as a hot reality it was exceedingly disagreeable.

The day was delightful, and we floated with such rapidity, that the quick succession of turns, and points, and islands, made time pass most pleasantly. We could readily imagine what a fairy scene the river would be could we pass with steamboat speed.

We longed to know what sort of arrangements Noah made for his parrots. Thus far, ours had been left pretty much to their own discretion, and the necessity for an immediate "setting up of family government," was hourly more urgent. The macaw, no wise contented with his elevation, had climbed down, and was perpetually quarreling with a pair of green parrots, and, all the time, so hoarsely screaming, that we were tempted to twist his neck. The parrots had to have a pitched battle over every ear of corn, and both they, and the macaw, had repeatedly flown into the water, where they but narrowly escaped a grave. There were two green paroquets and one odd one, prettiest of all, with a yellow top, and they could not agree any better than their elders. Yellow-top prided himself on his strength, and considered himself as good as a dozen green ones, while they resented his impudence, and scolded away, in ear-piercing tones that made the cabin an inferno. At other times, they all three banded together, and trotting about deck, insulted the parrots with their impertinences. When a flock of their relations passed over, the whole family set up a scream, which might have been heard by all the birds within a league; and if a duck flew by, which was very often, our geese would call in tones like a trumpet, and the guan would shrilly whistle. When we came to the shore, we were obliged to shut up our protégées in the tolda, or they were sure to scramble up the nearest limb, or fly into the water, and swim for the bank. Really, it would have troubled a Job; but we could see no relief.

In the afternoon, instead of taking a smaller passage, by which we had ascended, we continued with the main current, and passed a collection of houses, known as Tabocal. Each house stood upon a little point, overhanging the water; and the general appearance was neat and pleasing. The people were all fishermen, and the river, aided by a little patch of mandioca, supplied all their wants. There were, also, a great many orange trees, which indicated rather more providence than usual in the river settlers.

We shot a female Snake-bird, Plotus anhinga, in full plumage. The Indians asserted, very positively, that this was a different species from that found below, calling it, by way of distinction, the Carará de Rio Branco. We had no opportunity, afterwards, of verifying their account, and the only specimen that we had shot, upon our ascent, was a young male of this same species. But whether there be one species or two, the Darter is common every where upon the river, and upon Marajo. The Surinam Darter is probably quite as abundant, but from its small size, more easily overlooked. We obtained one of these at Barra, and, afterwards, saw several in a collection, at Jungcal.

Upon the 31st, as we were stopping in the forest to breakfast, our geese called up a kindred wild one, which we shot and preserved. This species I have before mentioned as the Chenalopix jubatus (Spix). It is more elegant in its movements than any of its family with which we are acquainted, being small, with long neck and legs, and extremely active. It walks with stately step, but usually its motion approaches a run, with outspread wings, and proudly arching neck. It is not seen at Pará, but is common above, and is much prized, by gentlemen, as ornamental to their yards.

At about ten o'clock, we reached the place, where, in ascending, we had seen a few Herons' nests. Now, the trees along the shore, were white with the birds; and a boat, moored to the bank, indicated that some persons were collecting eggs. Taking one of the men, with the montaria, leaving the galliota to float with the current, we started for the spot. The trees were of the loftiest height, and in every fork of the branches, where a nest could be formed, sat the female birds, some, with their long plumes hanging down, like the first curving of a tiny cascade; others, in the ragged plumage of the molting season. The male birds were scattered over the tree tops, some, hoarsely talking to their mates, others, busily engaged in dressing their snowy robes, and others, quietly dozing. The loud clamor of their mingled voices so deafened us, that we were obliged to speak to each to each other in screams. The report of the gun made no impression upon the thousands around, and the marked bird fell unnoticed. Many of the trees were half denuded of their bark, by the animals who had climbed up, and the tracks of tigers, large and small, exposed the marauders. We shot an iguana who was sucking the eggs from a nest, and the Indians, whom we found, assured us that they had seen large snakes in the trees on like errands. Dead birds strewed the ground, some partly devoured, and others nothing but skeletons, upon which the swarms of ants had feasted. Soiled plumes were in profusion, but ruined beyond redemption, and we did not care to gather them. There was to be seen but one pair of the Great Blue Herons, the rest were all the Great White Herons, A. alba. We shot about a dozen of these in fullest plumage, and prepared to hasten after our boat. There were two men collecting eggs, but owing to the size and loftiness of the trees, and the multitudes of stinging ants which infested them, they had made but little progress. They had ascended but one tree, and with a bag and string, had let down thirty-four eggs, which we bought for twelve cents. They were blue, and the size of small hens' eggs.

There was another breeding place of this kind opposite Serpa, and we had intended spending a day within it, had Sr. Manoel Jochin been at home.

We arrived at Villa Nova, about noon of August 1st, having in forty-eight hours made a distance, which required eight days in ascending. Senhor Bentos invited us to make his house our home during our stay, and we, at once, moved into it, leaving the galliota in charge of Pedro and his comrade. The commandant was absent, and we were likely to be detained some days, as no spare men were in the place, and several other voyagers were in the same predicament as ourselves. But there was no use in complaining, and come what might, we were in comfortable quarters.

When we went up, the town was crowded from the sitios in the vicinity, on account of the fiesta of St. Juan; but now, many of the houses were closed, their inmates being in the country, for the summer, and every thing bore an aspect of dreariness.

The next day was Sunday, but there were no services in the church, the Padre being absent on some of his trading expeditions; but in the afternoon, there was a procession of the women and children, preceded by "that same old" drum.

The commandant had returned, and we called to pay him our respects, and make known our wants. He was a very young man, and appeared anxious to oblige us by every means. in his power. He promised to forward us, with twelve men and a pilot, if we would only wait a few days, until he could obtain them from the woods. Of course, we could but choose the only alternative, though our friend's promise enabled us to bear the infliction with a tolerable grace. He was very indignant at the recital of our desertion by two of his men, and before he had heard the story out, had ordered them to the calaboose, with the etceteras.

This day was memorable, in that we then, for the first time since we had been in Brazil, saw tomatoes. They were little and few, for the climate is unfavorable to their growth. Okra is much more common, and is eaten both in soups and with boiled dishes. It seems strange that, directly under the equator, the Brazilians can live as they do, upon turtle, and meat, and fish. With all this, they consume vast quantities of cashaça, which is as bad as New England rum, and sleep, in the interior towns, about sixteen hours out of the twenty-four. And yet, we saw very many old men, of sixty and seventy years, and scarcely ever knew a case of sickness.

Next morning, a large party of us went to the lake. A well beaten road led to its side, and we found it a pretty sheet of clear water, in a valley of considerable depression. Large fields of grass were floating upon the surface, at the will of the winds, and from them were startled many ducks, Anas autumnalis, of which we shot enough for a dinner. They were now in pairs, just about to commence their breeding season; at which time, they resort to inland lakes, whither every one, who can raise a gun and a montaria, follows them. There were several Indian houses about this lake, and at a distance, were two men in montarias, engaged in taking Periecu. Every man of consequence, in Villa Nova, employs an Indian or black in fishing, selling the surplus of what he himself wants.

The Indians were building one of their largest vessels upon the beach at Villa Nova, and it was a matter of astonishment to us, that their carpenters could cut the planks and timbers with so great facility, and fit them with such precision, using only a handsaw, and the little adz of the country; while the timber was of almost iron hardness, and impenetrable to worms or insects. The shape of these river embarcaçoens is calculated for any thing but speed, they being broad, round-bottomed, and nearly square-bowed. A vessel after the model of the Hudson river sloop, would ascend the Amazon in half the time now required.

The little montarias are constructed in a different manner from Indian canoes in other countries. A log is selected, not more than a foot in diameter, and properly hollowed, through as narrow an aperture as will allow of working. This finished, it is laid over a fire, bottom side up, and the aperture is thus enlarged as is requisite. The outside is properly modeled, and upon either gunwale is fastened a strip of board, six inches in width, meeting at each end of the boat. They are usually about fifteen feet in length, and a load of Indian will cross the river, when the adges of their tottleish craft are scarcely above the water, and when white men would certainly be overturned. In such labor as boat building, timber hewing, paddling, and making of hammocks, the Indians enjoy an uncontested superiority, although in any other, they are worse than useless.

Our boatmen were to have arrived on Tuesday night, but upon going to the beach, the next morning, we saw the Commandant just pushing off, with eleven men in two boats. His sergeant, he said, had returned without a man, and he had ordered him to the calaboose, for disobeying orders; now he was going upon our errand himself, and would have the men at any rate. This Commandant was a noble follow, and, although he was acting under orders, yet he entered into our plans with so much good will, as to make us personally indebted to him. He had taken all the workmen from the boat, and the beach and town were as still as a New England village on a Sunday.

The poor sergeant, who was in durance for his misfortune, had the best reason in the world for not bringing the men, the first and most important point being to find them. This was no easy matter, when the hunted ones were unwilling Indians, in their own woods.

The military officers, in these inland towns, are despotic for evil or good, and according as they are public spirited men, does the town prosper. At Serpa, every thing appeared careless and disorderly; at Villa Nova, on the contrary, a change was evidently taking place for the better, and even since we had passed up the river, the vicinity had undergone an entire transformation. The soldiers had been employed in cutting down the bushes that encroached upon the town, in pulling down and removing the crazy hovels, in building handsome fences about the houses of the officers, and in clearing and repairing the road leading to the lake.

Near our house, a school was in daily session, and as the path to the woods ran directly by it, we took frequent peeps at the little fellows within. The master was a deputy, a boy of sixteen, and a flock of children, of all colors, were gathered around him, all talking or studying at the top of their voices. Here these future statesmen learned reading, and writing, and a little arithmetic. The Brazilians, generally, are very neat in their chirography. The government pays the salary of the head teacher, or Professor, as he is styled. In Villa Nova, his salary was one hundred and fifty milrees annually, from which he deputized as cheaply as possible. This Professor, Senhor Amarelles, who, by the way, was one of the dignitaries of the place, concentrating in himself some half a dozen offices, chanced to be in possession of a counterfeit note; and this, he desired the shopkeeper of the place to palm off upon us, as we, being strangers, he said, would not know the difference. Very dubious morality, for a schoolmaster.

Apropos, there were an unusual number of Vultures about Villa Nova, the Cathartes atratus of Wilson; and indeed, this species is seen more or less, every where upon the river. At Pará, particularly, they are seen by hundreds, about the slaughter yard, and with them may occasionally be seen a red-headed species, which we supposed to be the common Turkey Buzzard of the North, C. aura, but which, it has been suggested, may more probably be the Cathartes Burrovianus of Cassin. Unfortunately, we did not preserve specimens of this bird. There is a third species, the King of the Vultures, Sarcoramphus papa, or as it is called in Brazil, Urubu-tinga. The termination tinga, in the Lingoa Geral, means king, and this bird well deserves the name, from its beauty and superior strength. If a King Vulture makes its appearance where a number of the other species are collected about carrion, the latter instinctively give way, and stand meekly around, while their sovereign leisurely gorges himself. These birds are not very common upon the Amazon, and we never had an opportunity of shooting them, but several times, we observed them circling, in pairs, over the forest. Senhor Henriquez informed as, at the Barra, that they were not unfrequently taken alive, particularly if a putrid snake, of which they are fond, be exposed to them. A noose is arranged to fall over their heads, and the caught bird is transformed from a wild marauder, into a peaceable citizen. At Pará, they are highly valued. We saw a pair in perfect plumage, which were presented to Mr. Norris, and felt nothing of the disgust inspired by the other, common species. Their bare necks were beautifully marked with red and black, orange and yellow, and were surrounded near the base by a ruffle of feathers. Their breasts were white, and the general color of the upper parts, was a light ashy gray. These birds were very active, moving about the yard with a leap rather than a step.

At last, upon Saturday, the 8th, the Commandant returned successful, and by five o'clock in the afternoon we were ready to bid a glad adieu to Villa Nova. During our stay, Senhor Bentos had been perpetually studying ways of obliging us, and, at last, he overwhelmed us with all kinds of gifts, even to a hammock and towels. He killed a cow for us, packed up two baskets of chickens, sent down a pair of his pet land turtles, a supply of farinha and oranges, bought or begged a curious parrot from the Rio Tapajos, and added to it all the parrots which he had about the house, and even a basket of half-fledged doves. Moreover, after we had pushed from the shore, and descended several miles, a montaria overtook us with one of the Senhor's house servants, whom he had sent with orders to accompany us, as far as we wished, and to attend to our cooking. When the hour for parting came, we found the good old man in his hammock, the tears coursing down his cheeks, and apparently in great distress. He threw his arms about our necks, and sobbed like a child, and it was only after an interval of several minutes, that he let us go, loaded with a hundred blessing.

Our men were nearly all of the tribe of Gentios, the best upon the river. Among them were two free negroes, who had been admitted to the rights of tribeship. To look after them, the Commandant sent also a corporal and a sergeant; the former of whom was to be pilot, and the latter, a gentleman of leisure.

During the preceding night, Pedro had been seduced away by a white man, who was engaged in fishing, in some of the lakes. Pedro had seen quite enough of civilization, and longed for his woods and freedom again. We had found him one of the best natured fellows in the world, and there was no fault in him, except his inquisitiveness, which was natural enough. He was always for trying on our hats, or using our brushes and combs, or some similar liberty, and there was no use in attempting to explain the impropriety of the thing.

Our load was now considerably increased. The few turtle with which we had started from Barra, were reinforced to the number of fifteen, and filled all the space beneath the cabin floor, and a good share of the tolda. In the bow, some trader had stowed several pots of balsam, and had had the assurance to further impose upon our good will, by demanding a receipt for the same, which he did not get.

Early in the morning of the 10th, we passed Obidos. Sailing as we did, in the middle of the channel, the shores appeared to fine advantage, and yet, we could obtain but a very indifferent idea of the country, or of its productions, at such a distance. We had hoped to collect a number of birds and plants, whose localities we had marked in ascending, but we found it impossible to stop, even could we have recognized the proper places. We could only take counsel for the future, and resolve, that if ever we enjoyed another similar opportunity, we would not thus defer increasing our collection to a more convenient season.

Towards night, we stopped at the same high point, at which we had breakfasted, the second morning from Santarem. Now we were distant but six hours from that place. Here, by the deserted house, we found an abundance of oranges and limes. We shot a Caracara Eagle, Polyborus Braziliensis, a bird interesting to us, from its being also a resident of the United States. The Indians called it the Caracara Gavion. It is one of the smaller eagles, and somewhat allied to the vultures. We had often seen them, sitting upon trees not far from the water, and they seemed little shy at our advance. We afterwards saw them on Marajo, and, undoubtedly, they are common throughout the whole country. The Hawk tribe of birds was always exceedingly numerous, many being beautifully marked, and of all sizes, down to a species smaller than our Sparrow-hawk. We had shot many varieties, and shot *at* as many more.

Our men required no urging, and we found a vast change from the lazy Muras. The sergeant regulated their hours of labor, and we were unconcerned passengers. They were all young, and more inclined to frolic than other Indians that we had seen.

The sergeant had with him a curious musical instrument. It consisted of a hollow reed six feet in length, in one end of which was fitted a smaller joint, extending a few inches. In this was a blowing hole; and from the whole affair, our amateur produced sounds much like those of a bugle, playing a number of simple tunes. The men passed half their time in singing, and two of them, who seemed to be leaders, often composed a burden of their own, of the wonders they expected to see in the city, to which the others joined in chorus.

We inquired of them the name of the Amazon in the Indian tongue. It was Pára-na-tinga, King of Waters.

Chapter XVIII

ARRIVE AT SANTAREM—NEGRO STEALING—PASS MONTE ALEGRE—STRONG WINDS—USEFULNESS OF THE SUN-BIRD—FAMILY GOVERNMENT—REFORMATION IN THE PAROQUETS—LOW SHORE—A CONGRESS—OTTERS—ENTER THE XINGU—GURUPÁ—LEAVE THE AMAZON—ASSAI PALMS—A FRIEND LOST AND A FRIEND GAINED—BRAVES—OUR WATER JARS—CROSSING THE BAY OF LIMOEIRO—SERINGA TREES—A LOST DAY—TOWN OF SANTA ANNA—IGARIPÉ MERIM—ENTER THE MOJU—MANUFACTURE OF RUBBER SHOES—ANATTO—ARRIVAL AT PARÁ

We arrived at Santarem about midnight, and anchored off the house of Captain Hislop, waiting for the morning. The Captain was absent, but had left orders to place his house at our disposal. Therefore, without further ceremony, we took possession, and breakfasted, once more, upon the delightful Santarem beef. We called upon our friend Senhor Louis, and were gratified to find that he had not forgotten us, in our absence, but had made for us a good collection of insects, and other matters, in which we were interested. He pressed us much to protract our stay, as did Mr. William Golding, an English resident, who called upon us; but our loss of time at Villa Nova obliged us to make all speed to Pará.

The large black monkey, which had been given us two months before, and whose society we had anticipated with mingled emotions, had gone by the board, about a week previous, "laying down and dying like a man," as the old lady said. To console our bereavement somewhat, she sent down to the galliota, a pair of young, noisy, half-fledged parrots, and a Pavon, or Sun-bird. Senhor Louis added a basket of young paroquets, and a pair of land turtles, and Mr. Golding a pretty maraca duck. Thus we were to have no lack of objects for sympathy or entertainment, for the remainder of our voyage.

We do not know how near we came to getting into difficulty with some of Santarem's officials, although innocent of all intention of offending. Senhor Bentos' servant had gone ashore, and called upon the sister of the Senhor; and, probably, not exactly understanding, herself, why she had been forwarded in our boat, had made an unintelligible story of the whole matter. The Senhora sent us a polite request to visit her, which we did; and to her inquiries, we answered as we could. She was anxious that we should see her brother-in-law, who could not call upon us, she observed, "because his neck was so short, and his belly so big," and offered to send a servant with us to the gentleman's house. We could not refuse, and went accordingly. The Senhor was in his hammock, and it was evident enough, that his sister's expression was truthful, at least, for he was sorely afflicted with dropsy. He was a lawyer, and after thanking us for our attention, commenced a legal cross-examination of the whys and wherefores of the wench's case. It was no joke to be suspected of negro stealing; but we replied according to our ability, that we had received no instructions from Senhor Bentos, that the woman had come on board without our wishing it, that she had stayed on board without our needing her services, and that we had brought her to Santarem, because we had not stopped elsewhere. Just at this time, came in a gentleman whom we had known at Pará, and after a few words of explanation, we were bowed out of the house with the profoundest civility. And we would advise no Amazon voyager to receive in charge negro cooks, unless their master comes with them.

We left Santarem as the sun was setting, and the men being favorably inclined, we made rapid speed during the night.

We passed Monte Alegre upon the afternoon of the next day, the 12th. It had been our intention to stop, for a few hours, at this town, for the purpose of obtaining specimens of the beautiful cuyas there made, and for a ramble upon the mountain in the vicinity. But a strong breeze drove us into the remoter channel, at least fifteen miles from the town, and we could not cross.

During the night, a furious wind, accompanied by rain, prevented our advance. Early upon the 13th, we stopped in a small bay, for a few hours, until the sea should abate. The men slung their hammocks under the trees, or stretched themselves on logs, as they could find opportunity. For ourselves, we got out the lines, and fished with decided success. We also shot a pair of geese, which were called up by our decoys.

At this spot, our cabin was filled with a large fly, the Mutúca, which, in the dry season, is almost as great a pest, by day, as the cárapaná by night. But here our Pavon showed himself useful, walking stealthily about the floor, and picking off fly after fly, with inevitable aim. Not many days after, we discovered that he was as fond of cockroaches as of flies; and it was then a regular pastime to put him in one of the lockers, and stir up the game, which we had no difficulty in finding, nor he in catching.

Our noisy additions from Santarem made longer endurance out of the question, and after long threatening, at last we succeeded in "setting up the family government." As the first overture thereto, a rope was crossed a few times in the tolda. Upon this, the arara and the parrots were placed, with the understanding that they might look out of the door as much as they pleased, and be invited thence, at regular hours, to their meals; but that further liberties were inadmissible and unattainable. So there they sat, scarcely knowing whether to laugh or cry. The paroquets were stationed at the afterpart of the cabin, and the change, which had come over one of the green ones, from Barra, was amusing. She had been the wildest, and crossest little body on board, always resenting favors, and biting kindly hands. But since the lately received young ones had been put with her, she had assumed all the watchfulness of a mother, feeding them, taking hold of their bills and shaking them up, to promote digestion, and generally keeping them in decent order. She had no more time to gad about deck, but soberly inclined, with the feathers of her head erect and matronly, she stuck to her corner, and minded her own business. Meanwhile, Yellow-top looked on with the calm dignity of a gentleman of family.

When opposite Pryinha, we took an igaripé, to avoid the long circuit and the rough channel, and sailed many miles upon water, still as a lake. Here were vast numbers of ducks and ciganas, Opisthocomus cristatus. These latter had lately nested, and the young birds were in half plumage. They seemed to be feeding upon pacovas, which grow in abundance upon the grounds of a deserted sitio; and as we startled them, they flew with a loud rustling of their wings, like a commotion of leaves, hoarsely crying, *cra, cra.* The nests of these birds, are built in low bushes; and are compactly formed of sticks, with a lining of leaves. The eggs are three or four, almost oblong, and of a cream color, marked with blotches of red and faint brown.

During the night, the wind blew with such strength, as to drive us towards shore; and several times, we were among the cárapanás, or running up stream in the romanças, almost equally disagreeable.

Where we stopped, next morning, the 14th, the whole region had been overflowed, upon our ascent. Now, the waters had fallen three feet; and the land was high and dry, and covered by a beautiful forest. While at this place, extraordinary noises from a flock of parrots, at a little distance, attracted our attention. At one instant, all was hushed; then broke forth a perfect babel of screams, suggestive of the clamor of a flock of crows and jays about a helpless owl. It might be, that the parrots had beleaguered one of these sun-blinded enemies; or, perhaps, the assembly had met to canvass some momentous point, the overbearing conduct of the araras, or the growing insolence of the paroquets. Guns in hand, we crept silently towards them, and soon discovered the cause of the excitement. Conspicuously mounted upon a tree top, stood a large green parrot, while around him, upon adjacent branches, were collected a host of his compeers. There was a pause. "O Jesu ———u," came down from the tree top; and a burst of imitative shrieks and vociferous applause followed. "Ha, ha, ha ———a," and Poll rolled his head, and doubled up his body, quite beside himself with laughter. Tumultuous applause, and encores. "Ha, ha, ha, Papaguyri ———a," and he spread his wings and began to dance on his perch with emphasis. The effect upon the auditory was prodigious, and all sorts of rapturous contortions were testifying their intelligence, when some suspicious eye spied our hiding place; and the affrighted birds hurried off, their borrowed notes of joy ludicrously changed to natural cries of alarm. Complacent Poll! he had escaped from confinement; and with his stock of Portuguese, was founding a new school among the parrots.

In the afternoon, we entered the igaripé, through which we had sailed upon the 11th of June, occupying then, the entire day; but which now required but two hours. Here we saw a number of Otters. The men called them by some wild note; and, immediately, the animals raised their heads and shoulders above the surface of the water, and listened without the least, apparent fear. It was almost too bad to spoil their sport; but the opportunity was too tempting, and straightway, amongst them whizzed a ball. They dived below, and we saw them no more.

When ascending, we had seen the mountains, upon the northern side of the river, for several days; but, as we left this igaripé, they broke upon us, in one full view, seemingly of twice the height, and tenfold the beauty, of the mountains we had seen before.

Next morning, the shore was very low; scarcely dry from the receding waters. A mud flat extended for more than a mile into the river; and the top of the water was spotted by roots and stumps of trees.

Towards night, we left the Amazon, for a narrow passage, which led into the River Xingu; and, for several hours, our course was in the clear waters of that river, among islands of small size and surpassing beauty. Just at sunset, as we were proceeding silently, there came floating over the water, the rich, flute-like notes of some evening bird. It was exactly the song of the Wood Thrush, so favorite a bird at the North; and every intonation came freighted with memories of home, of dear ones, far, far away. Even the Indians seemed struck with an unusual interest, and rested upon their paddles to listen. We never had heard it before; and so strangely in unison was the melody with the hour and the scene, that it might well have seemed to them the voice of the "spirit bird." We passed the small town of Boa Vista. At first, there seemed to be but one house, from the light; but the noise of our singing attracted attention, and a dozen torches welcomed us to shore, if we would.

Here we bad first made the acquaintance of the cárapanás, and here we left them for ever. They had clustered around us in prosperity and adversity, with a constancy that might have won the hearts of those, who were stronger nerved, or whose sympathies were more expanded than ours; but we parted from them in ungrateful exultation.

We reached Gurupá, about noon, of the 16th. Here we first received tidings of the war between the United States and Mexico. *Seventy thousand volunteers*, our informant said, had passed over the Mexican frontiers, and were advancing, by rapid marches, to the borders of Guatemala!

It was three o'clock, the next afternoon, when we stood upon the cabin-top, for a last look at the main Amazon; and as a turn of the Tajipuru, into which we had now entered, shut it suddenly from our view, we could not but feel a sadness, as when one parts from a loved friend, whom he may never see more. The months that we had passed upon its waters, were bright spots in our lives.

Familiarity with the vastness of its size, the majesty, and the beauty of its borders, the loveliness of its islands, had not weakened our first impressions. He was always the King of Rivers, stretching his dominions over remotest territories, and receiving tribute from countless streams; moving onward with solemn and awful slowness, and going forth to battle with the Sea, in a manner befitting the loftiness of his designation, and the dignity of his claims.

We were now sailing, in narrow channels, towards Braves; but by a different route from that of our ascent. A great number of channels, from the Amazon, intersected our course, through which the water poured furiously. The shores again bristled with palm trees; or forests of seringa, and the huts of the gum-collectors, skirted the stream.

We gathered great quantities of assai, and, ourselves turning artists, we could have it in Pará perfection, and could bid adieu without a thought, to our stores of coffee and other former indispensables, which were disappearing, one after another; a sure token, that, by this time, our voyage should have ended.

Our motherly Paroquet came upon deck, for an airing, and embraced the opportunity of a high starting point, and a near shore, to give us French leave; but, a few hours after, as if to supply her loss, we picked up a little Musk Duck, not more than a day or two from the shell. Tile little fellow was all alone, his mother having taken flight, at our approach, and his brothers and sisters, very likely, having fallen prey to some water enemy. He was wild enough at first, but soon became extremely familiar, and was the pet of the cabin. Now, he swims, in matured and beautiful plumage, in one of our New York ponds, and, we trust, that when his flesh returns its dust to dust, it will be when his head is gray, and his years honored, and without the intervention of Thanksgiving epicure, or Christmas knife.

Late in the evening of the 18th, we reached Braves, the same little old town that we had left it. We went on shore for our much desired water-jars, and found that the old woman had fulfilled her promise, for there they stood, glazed and finished, amongst a row of gaudy brothers, that quite looked them out of countenance. We offered to pay for them in two milree notes, which, being at a slight discount, were not received. Then we offered Spanish dollars, but the jackass of a storekeeper did not exactly like the appearance of those bright-looking things, and refused to receive any thing but copper. We had no copper, and came away, with a hearty, and heartily expressed wish that the jars might stand upon his shelves till his head was gray.

Leaving Braves, with the morning tide in a few hours, we had passed out of the narrow channels, and were fairly crossing the Bay of Limoeiro, taking what is called the Cametá route, the usual one for vessels bound down. For three days we were crossing from one island to another, often twelve and fifteen miles apart, and in what looked more like a sea, than the mouth of a river. The channel was not very distinct, and our pilot knew little of his business. Every where were shoal banks, exposed at low tide, and, many times, we struck upon the bottom, which, fortunately, was no harder than mud.

The men were growing eager for the city, and soon after midnight, upon the morning of the 22^d, they started, of their own accord, and for a couple of hours, we went on swimmingly. But a strong wind arose, and the rising waves tossed our frail boat somewhat uncomfortably. For some hours, we coasted along a sand bank, in vain endeavoring to attain a passage to the island, an hundred yards within, frequently striking with such violence as to make us fearful that the bottom of the boat would be stove in. At last, about daybreak, we contrived to set two poles firmly in the mud, and tying our boat to them, we were pitched and rolled about, as if in an ocean storm. The men swam to shore, and caught a breakfast of shrimps, in pools left by the tide. Towards noon, as the flood came in, we were able to moor nearer the trees, and beyond reach of the wind.

This island was covered by a fine forest, in which were abundance of Seringa trees, all scarred with wounds. We made some incisions, with our tresádos, and the milk, at once, oozed out, and dripped in little streams. Its taste was agreeable, much like sweetened cream, which it resembled in color. These trees were, often, of great height, and from two to three feet in diameter. The trunks were round and straight, and the bark of a light color, and not very smooth. The wood was soft, and we easily cut off a large root, which we brought away with us. The top of the Seringa is not very wide spreading, but beautiful from its long leaves, which grow in clusters of three together, and are of oblong-ovate shape, the centre one rather more than a foot in length, the others a little shorter. These leaves are thin, and resemble, in no respect, the leaves of an East-Indian plant, often seen in our green-houses, and called the Caoutchouc. There is not, probably, a true Seringa in the United States. Around these trees were many of the shells

(Ampullarias), used in dipping the gum, and also, some of the mud cups, holding about half a gill each, which are fastened to the tree, for the purpose of catching the gum, as it oozes from the wound. We found also the fruit of the Seringa. It is ligneous, the size of a large peach, divided into three lobes, each of which contains a small black nut. These are eagerly sought by animals, and although the ground was strewed with fragments, it was with great difficulty that we found a pair in good preservation. Specimens of all these things, wood, leaves, shells, cups, and seeds, we secured. The manufacture of the gum we had not yet seen, but shall describe shortly.

The waves somewhat subsiding, and the wind being more favorable, we started again at two in the afternoon, this being our last crossing. The point, at which we aimed, was about fifteen miles distant, and we arrived near the shore, soon after sundown. But here, we were again entangled in shoals, and for a long time, were obliged to beat backwards and forwards, endeavoring to find the channel, with the comfortable feeling to incite us, that the tide was rapidly running out, and that we bade fair to be left high and dry in the mud. At last, we found the right course, and were soon stopping at a house, at the entrance of an igaripé. Here, we were told that our passage had been very perilous, and that, only the day before, a vessel, loaded with cacao, had gone to pieces, upon these same shoals. We engaged a man to go with us, to pilot our pilot, and starting once more, pulled all night.

The morning of the 23rd, found us in a narrow stream, and soon after sunrise, we stopped at a deserted sitio to breakfast. Here our guide left us, returning in his montaria, as our pilot declared; that now he perfectly remembered the way. We sailed on, the streams winding about in every direction, and passed many sitios, and sugar *engenhos*, upon the banks. At eleven o'clock, we came to a very large house, which our pilot said was that of the Delegarde of Santa Anna, and that now that town was but two turns ahead. We continued on two turns, and twenty-two turns, but without seeing the lost town, although our necks were strained, and eyes weak, with the search. As fortune would have it, a montaria came down the stream, and we learned, to our dismay, that we were in the river. Murué, altogether the wrong stream, and that we had deviated from the main and evident course, soon after breakfast. Moreover, that, had we not chanced to meet this montaria, we might have gone on, all night, through the forest, without seeing a house, or a man. Here was the time for all our philosophy. Turning back, after a few hours, we struck into a cross stream, and at last, were in the Kixi, the river upon which Santa Anna stands. It was midnight, when we arrived at this town. It is an excise port, and every vessel passing, pays a toll of ten vintens. We were hailed by a guard and ordered to stop. Our sergeant had put on his uniform, and now went on shore to adjust matters; while we remained, viewing the town, as we could by starlight. Starlight undoubtedly flatters; still, Santa Anna is considered the prettiest little town in the province. A large church, of fine proportions, stands directly by the shore; the houses are well proportioned, and good looking; and, fronting the stores, are wharves built out into the water. The town derives much of its importance from its being a port of excise; but all the surrounding country is thickly settled by sugar planters and growers of cotton.

The sergeant, returning, reported no duties, as he had told the officer that we were public business, bearing his majesty's mail.

Between Santa Anna and the river Mojú, is the igaripé Merim, a short canal cut through by government, for the purpose of enabling vessels to reach Pará more readily, and to avoid a tedious circuit. Striking into this, we continued down with the tide, and daybreak of the 24th found us far advanced upon the Mojú. This is small stream, and its banks are covered with flourishing plantations. We passed what appeared to be the ruins of a village, consisting of a large church, and a few houses.

At ten o'clock, we stopped at an anatto plantation, awaiting the tide, and here we saw the manufacture of rubber. The man of the house returned from the forest about noon, bringing in nearly two gallons of milk, which he bad been engaged, since daylight, in collecting from one hundred and twenty trees, that had been tapped upon the previous morning. This quantity of milk, he said, would suffice for ten pairs of shoes, and when he himself attended to the trees, he could collect the same quantity, every morning, for several months. But his girls could only collect from seventy trees. The Seringa trees do not usually grow thickly, and such a number may require a circuit of several miles. In making the shoes, two girls were the artistes, in a little thatched hut, which had no opening but the door. From an inverted water jar, the bottom of which had been broken out for the purpose, issued a column of dense, white smoke, from the burning of a species of palm nut, and which so filled the hut, that we could scarcely see the inmates. The lasts used were of wood, exported from the United States, and were smeared with clay, to prevent adhesion. In the leg of each, was a long stick, serving as a handle. The last was dipped into the milk, and immediately held over the smoke, which, without much discoloring, dried the surface at once. It was then redipped, and the process was repeated a dozen times, until the shoe was of sufficient thickness, care being taken to give a greater number of coatings to the bottom. The whole operation from the smearing of the last, to placing the finished shoe in the sun, required less than five minutes. The shoe was now of a slightly more yellowish hue than the liquid milk, but in the course of a few hours, it became of a reddish brown. After an exposure of twenty-four hours, it is figured, as we see upon the imported shoes. This is done by the girls, with small sticks of hard wood, or the needle-like spines of some of the palms. Stamping has been tried, but without success. The shoe is now cut from the last, and is ready for sale: bringing a price of from ten to twelve vintens, or cents, per pair. It is a long time before they assume the black hue. Brought to the city, they are assorted, the best being laid aside for exportation as shoes, the others as waste rubber. The proper designation for this latter, in which are included bottles, sheets, and any other form excepting selected shoes, is *boráçha*, and this is shipped in bulk. There are a number of persons in the city, who make a business of filling shoes with rice chaff and hay, previous to their being packed in boxes. They are generally fashioned into better shape by being stretched upon lasts after they arrive at their final destination. By far the greater part of the rubber exported from Pará, goes to the United States, the European consumption being comparatively very small.

At this place, we found the largest and finest oranges that we had ever seen, and, for about twelve cents, purchased a bushel.

Anatto is a common product in the vicinity of Pará, but in no place is it cultivated to much extent. The plant is the Bixa Orellana. It is a shrub, growing much like the lilac, and bears a dark leaf, similarly shaped, but much larger. The clusters of fruit pods contain numerous small red seeds, which yield the substance known as the anatto of commerce, and which is used extensively in coloring cheese. It is difficult to obtain the anatto in a pure state. Its color so much resembles that of red clay, as to render adulteration easy and profitable.

Late in the evening we arrived at Jáguary, the place of the late Baron Pombo, who was the greatest proprietor in the province, owning more than one thousand slaves, and cultivating an immense territory. The village consists almost entirely of the residences of those dependent upon the estate; and the bright light of torches, and the noise of various factories and mills, indicated that labor was exerting itself by night as well as by day. We moored close under the Baron's house, a large, pelace-like edifice.

Starting once more, at two in the morning, of the 25th, by three we had crossed the Acará, and by daybreak, were within sight of the city. The music of the band, the ringing of the bells, and the distant hum, came towards us like water to thirsty souls. The men broke out into a joyous song, and with a lively striking of their paddles, beating time to their quick music, they sped us past canoe after canoe, that, in easy indolence, was coursing like ourselves.

At eight o'clock, we were once more upon the Punto da Pedras, the spot we had left one hundred days before, receiving the warm congratulations of friends, and the curious attentions of a motley crowd, who had collected to gaze at the strangers from the Sertoen.

Chapter XIX

***OUR LADY OF NAZARETH—NAZARÉ LEGEND—
PROCESSION—COMMENCEMENT OF THE FIESTA—A
WALK TO NAZARÉ—GAMBLING—SERVICES IN THE
CHAPEL—AN INTERESTING INCIDENT***

Shortly after our return, commenced the festival of Nazaré.

This is the grand holiday of Pará, when business is suspended, and citizens have no care but pleasure. Our Lady of Nazareth seems to have received proper honors of old, in the mother country, and the faithful colonists still acknowledged her maternal kindness by enshrining her as their most popular tutelary. Did trouble afflict, or sorrow bow down; did danger menace, or were dangers escaped, our blessed Lady was ever considered the friend and benefactress. Many are the traditions of her miraculous interpositions and wonderful cures, all tending to prove how well she deserves the exalted place she holds in the hearts of all good citizens.

Befitting so beneficent a Saint is the beautiful spot devoted to her worship; a neat chapel within an ever verdant forest-embowered meadow. Quite lately, a number of graceful cottages have been erected about the area, mostly by wealthy persons in the city, who prefer to live here during the fiesta. At this time, numerous temporary constructions also line the adjacent road on either side, or find room about the square. The time usually chosen, by long custom, is the last of September, or early in October, when the increasing moon throws her splendors over the scene, and the dry season has fairly ushered in the unclouded, brilliant nights; when the air is redolent of perfume, and delicious coolness invites from the closeness of the city.

Associated with the kind offices of our Lady is an ancient legend, deemed worthy an annual recollection. It is of a knight, who, when rushing over an unnoticed precipice in pursuit of a deer, was saved from destruction by the timely apparition of our Lady, which caused the deflection of his affrighted horse.

It was about four in the afternoon, when the fierce sun's heat began to lose its power, that the procession which was to commence the fiesta by escorting our Lady to her chapel, formed in the Largo da Palacio. Amid the din of music, the discharge of rockets, and the vociferous applause of a vast crowd of blacks, it set forth. We had accepted the kind offer of a friend, and were watching from a balcony in the Rua da Cadeira. As the line approached, first and most conspicuous was a car drawn by oxen, in which were stationed boys having a supply of rockets, which at little intervals they discharged. Nothing so pleases a Brazilian as noise, especially the noise of gunpowder; and not only are rockets crackling night and day upon every public occasion, but the citizens are wont to celebrate their own private rejoicings by the same token.

Directly behind this car came another, similarly drawn, upon which was a rude representation of the before mentioned legend;—a monster of a man upon a caricature of a horse, being about to leap into space, while a canvass virgin upon the edge of the rock, or rather in the middle of the cart, prevented the catastrophe. Behind her was an exquisite little deer, no canvass abomination, but a darling of a thing, just from the forest, wild and startled. The poor thing could not comprehend the confusion, and would gladly have escaped, but the cord in its collar forced it back, and at last seeming resigned to its fate, it lay motionless upon its bed of hay.

Next followed the carriages, and therein, the pictures of complacence, sat the civic dignitaries and civic worthies. As locomotion is the sole object, every thing that can contribute thereto, from the crazy old tumble-down vehicle of the conquest, through every description of improvement, until the year '46, is pressed into the service. Most noticeable in this part of the procession is the President, a fine looking man, whose attention is constantly occupied by his fair friends in the balconies. Here and there, is a foreign consul, conspicuous among whom is the official or her Majesty of England, a venerable, soldierly figure, one of Wellington's campaigners and countrymen, and occupying decidedly the most dashing turnout of the day. Last of the carriages, comes a queer looking vehicle, known by no conventional name, but four-wheeled, and resembling the after part of an antique hackney coach, cut in two vertically and crosswise. In this sits a grave personage, holding in his hand the symbol of our Lady, to all appearance, a goodly sized wax doll, in full dress, magnificent in gaudy ribbons, and glowing with tinsel. Nossa Senhora is the darling of the crowd, and her attractions have lost none of their freshness during her year's seclusion.

Now come the equestrians, whose chargers do credit to their research, if not to the country which produced them; now and then one being a graceful animal, but the greater number, raw boned, broken-winded, down-hearted, and bat-bitten. After these, come black-robed priests, students in uniform, and genteel pedestrians, and, last of all, the military in force, preceded by their fine band.

Passing through the more important streets, the long line turns its course towards Nazaré, and here our Lady is deposited upon the altar of her chapel, and the fiesta has fairly begun.

The fiesta is of nine days' duration, and service is performed in the chapel every evening. For the first two or three days, the people are scarcely in the spirit of the thing, but before the novena is ended, the city is deserted, and its crowds are at home in Nazaré. Let us take a sunset walk, and see what is curious in a Pará festival. The brightness of day has passed with scarcely an interval, into the little inferior brilliance of the full moon. The trades, that blow more freshly at night, unite with the imperceptibly falling dew in exhilarating alter the day's fatigues. Lofty trees, and dense shrubs throw over us their rapidly varying shadows, and from their flower homes, the cicadas, and other night insects, chant their homage to the blessed Lady, in a vesper hymn. Grave matrons are passing along, attended by servants bearing prayer books; and comfortable looking old gentlemen, who have forgotten age in the universal gayety, are rivaling young beaux in the favors of laughing girls, whose uncovered tresses are flashing in the moonlight, and from whose lips the sweet tones of their beautiful language fall on the ear like music. Indians move silently about in strong contrast to the groups of blacks, the same noisy, careless beings, as elsewhere. Numbers of wenches, picturesquely attired, are bearing trays of doçes upon their heads; and children, of every age, add their share of life and glee to the scene.

Suddenly we leave the road, and the square is before us. The air is brilliant with torch lights; crowds of indistinct, moving figures are crossing in every direction, and the noisy rattle of a hundred gambling tables drowns all other sounds. These tables are as remote from the chapel as possible, and are licensed by the authorities. Upon each table are marked three colors, black, red, and yellow. The proprietor holds in his hand a large box, in which

WEDNESDAY
02/15/06
Senior

Museum Admission

1377708 493984 8055749 03:11pm 02/15/06 CM10

Enjoy your visit!
Be sure to visit the
Museum Shop.

32.00

WWW.AMNH.ORG

AMERICAN MUSEUM OF NATURAL HISTORY

02/15/06
3:30pm Butterflies
03:30pm   Senior

377708 493984 8055749 03:11pm 02/15/06 CM10   32.00 CA

American Museum of Natural History

Save $10 on a new family membership today!

Enjoy unlimited free general admission, discounts and more. Redeem this ticket today at any Membership Desk to receive a $10 discount on a **new** Family or higher level membership. For further information, call the Membership Office at 212.769.5606. Limit one ticket per membership. This offer may not be combined with any other offer. Valid only on date of purchase.

For ticket reservations call: 212.769.5200

Valid only for date and show time sold; no refunds or exchanges.

No food, drink, cameras or recording equipment allowed in theaters.

Management reserves the right to refuse admission to the holder of this ticket by refunding purchase price.

All seating by general admission; no late seating.

www.amnh.org

9574118

are a number of correspondingly colored balls. Whoever is inclined, stakes his money upon either color; a little door opens in the side of the box, a ball comes forth, and he has lost or won; probably the former, for the chances are two to one against him. But adverse chances make no difference, and crowds are constantly collected about the tables, mostly of little boys who have staked their last vinten, and who watch the exit of the ball with outstretched necks, starting eyes, and all the excitement of inveterate gamblers. It is amusing to watch these scenes. The complacent proprietor, very likely a black boy, grinning so knowingly at the increasing pile before him, and at the eagerness of his dupes, is evidently in sunshine. The poor little fellow who has lost his all, turns away silently, with dejected look, and tearful eyes. But let him win. A proud satisfaction brightens up his face, he looks around upon his unsuccessful mates with an air of most provoking triumph, and slowly rakes the coppers towards him, as though they could not be long enough in coming. Sometimes a pretty Indian girl hesitatingly stakes her treasure, timidly hoping that she may yet be the fortunate possessor of some coveted trinket: but, alas, the divinities here are heedless of black eyes and raven hair, and she turns away disappointed. At another stand, nothing less than paper is the etiquette, and some of Pará's bucks seem inclined to break the bank or lose their last milree.

Scattered every where over the square, are the stands of the doçe girls, who are doing a profitable business. Some of the cottages round about are fitted up with a tempting display of fancy wares; others are used as cafés, or as exhibition rooms for various shows; and from others come the sounds of music and dancing. Ladies and gentlemen are promenading about, waiting the commencement of the ceremonies in the chapel.

In all this crowd there is perfect order, and no drunken brawl or noisy tumult demands the police.

At eight o'clock, service is notified by the ascent of rockets, and those who care, attend the chapel. Within are the more fashionable ladies, and a few gentlemen; without, in the large open portico, are seated upon the floor the black and Indian women, dressed in white, with flowers in their hair, and profusely scented with vanilla. The congregation is still, the ceremonies proceed. Suddenly a sweet chant is commenced by the choir, one of the beautiful Portuguese hymns. The chorus is caught by the crowd in the portico. An old negress rises upon her knees, and acts the part of chorister and guide; in a voice almost drowning the sweet tones about her, calling successively upon all the saints of the calendar. "Hail to thee, Santo Tomasio. Hail to thee, Santo Ignacio." Certainty, she has a good memory. There is something indescribably beautiful in the tones of these singers. Men, women, and children, all join in the same high key, and the effect is wild and startling.

The service is over, and the amusements succeeding encroach far into the small hours of morning. Balls and parties are given in the cottages, or beneath the broad spreading trees, and the light hearted and happy, dance, until they are weary, to the music of the guitar, or their own songs.

While we were in Pará, an interesting incident occurred to diversify the festival. A few weeks before, a Portuguese bark had left Pará for Lisbon. One day out of the river, in the early morning, a squall struck her, threw her upon her beam's end, and she was capsized before a single passenger could escape from the cabin. The mate and seven seamen were thrown unhurt into the water. The small boat was likewise cast loose, and this they succeeded in attaining. They were in the ocean, without one morsel to eat, or one drop of water. For several weary days they pulled, and worn out by hunger and thirst, they laid them down to die. They had implored the aid of our Lady of Nazareth, had made her a thousand vows, but she would not save them. One rises for one more last look; land is in view. Hope rouses their wasted frames, and they reach Cayenne in safety. The inhabitants succor them, and send them to Pará with the boat, whither they arrive during the fiesta, bringing the first accounts of the disaster. The enthusiasm of the people was extreme. An immense procession was formed. The boat was borne upon the shoulders of the saved men, and deposited with rejoicings in the portico of our Lady's chapel, another memorial of her kindly aid.

Chapter XX

Leave Pará for Marajo—Voyage—Cape Magoary—Islands—A morning scene—Arrive at Jungual—A breakfast—Birds—Vicinity of Jungcal

The far famed Island of Marajo, a little world of itself, differing from aught else in its appearance, its productions, its birds and its animals, had long been to us an object of the most intense curiosity. Did we inquire the whereabouts of any curious animal of the dealer in the Rua, almost invariably the answer was, Marajo; or the locum tenens of some equally curious bird, of the wenches on the Punto da Pedras, of course, it was Marajo. Could not we catch a glimpse of an alligator? Yes, thousands on Marajo. And monster snakes and tigers? Always on Marajo. One would have thought this island a general depot, a sort of Pantological Institute, where any curiosity might be satisfied by the going. Ever since we had been in the country, we had heard of it, had seen, occasionally, the distant tree tops, and had even coasted along its upper side in the galliota; but our longings for a face to face acquaintance, and an exploration of its wonders, seemed likely to remain ungratified. And yet, we had been upon the eve of seeing Marajo for the last thirty days, thanks to Mr. Campbell's kindness. But the fiesta of our Lady of Nazareth, and the slow and easy habits of the people had kept us waiting from day to day, until the Undine's arrival, and expected speedy return, bade us bend our thoughts homeward.

But our intention was fulfilled, after all. At an hour's notice, we left Pará, about nine o'clock, one pleasant evening in September, dropping down with the ebbing tide. Our destination was Jungcal, upon the remote north-west corner of the island. The distance is not very great; a clipper schooner would call it a holiday excursion, and a little steamer, which could mock at the trades and the flood tides, would run it off in a pleasant morning. As it is, and alas, that it should be so, the Jungcal passengers think themselves fortunate, if the winds and tides of a week speed them to the destined point. Our craft was a cattle boat, a little schooner without a keel; with the least possible quarter-deck, and scanty turnings-in for two, below. A year before, we should have quarreled with the rats and cockroaches, but our recent experience had endued us with a most comfortable coolness in our manner of taking such small inconveniences. The crew were half-breeds, about a dozen in all, men and boys. The captain was a mulatto, not over twenty years of age, intelligent and sufficiently attentive. Had it not been for these attractive qualities, we should have grumbled unconscionably at a speculation of his, whereby, to deposit an Indian woman, who had ventured on board as passenger, in the steerage, he had lost an entire day in crossing to the Marajo side and back again. One would naturally suppose, that once upon the island shore, we could have coasted around Cape Magoary without re-crossing. But the river is beset with shoals, and no careful survey has yet sufficed to put these mariners at their ease.

Early upon the fourth morning, we struck across from Point Taipú, sixty miles only below Pará, and soon were running towards Cape Magoary with no guide but the stars, beyond view of land on either side. Our careful captain himself took the helm, and as we neared the shoals a man was constantly heaving the lead. The channel now was usually but one and two fathoms deep, and the brackish taste of the water was soon lost in the overpowering current which set in from the main Amazon. Beyond Cape Magoary are a number of small islands, the names of three of which are the Ship, the Bow, and the Flycatcher, or Navio, Arco, and Bentivee; all uninhabited by man, and affording secure homes to countless water birds. The isle of the Bow is overrun with wild hogs, the increase or a tame herd once wrecked upon a shoal near by. Here the captain offered to land us for an afternoon's sport, but the wind was fresh, and we were too near Jungcal for any such enticements. Late in the evening, we crossed the bar, passing into a small igaripé, and, in a few minutes, were moored off the cattle-pen. Once more we slept quietly, undisturbed by surfs and tossings.

The morning dawned in all the splendor of a tropical summer, and long before the sun's rays had gilded the tree tops, we were luxuriating in the fresh, invigorating breeze, and admiring the beautiful vicinity, that wanted not even the sunlight to enchant us. The ebbing tide had left exposed a large flat, extending an eighth of a mile opposite the cattle-pen, and lost, at perhaps, twice that distance, in the woods above. Here and there a tiny stream crept slowly down, as if loath to leave the beautiful quiet island for the rough waters beyond. Directly at our side, an impervious canebrake shot up its tasseled spires, rustling in the wind; while in every other direction, was piled the dark, massive foliage of tropical shrubs and trees. Above, and beyond reach of harm, a number of Great Blue Herons were stalking solemnly about, and near them, a company of Spoonbills and White Egrets displayed to us their delicate tints, in the increasing light. Opposite, a constantly gathering flock of large While Herons were intently watching our movements, as though balancing in their own minds the chances of danger, with the prospect of no breakfast, and a hungry family at home.

But the loveliest views will tire, in time, and despite the interest we felt in the position of things about us, when hour after hour passed away, and the gentle twilight became the fierce morning heat, while the scarce perceptibly ebbing tide would in no wise speed its movements in our behalf, we began to feel somewhat like prisoners, in durance. So, to vary the scene, we ventured, by the kindly aid of some tottering poles, to gain the shore, and started to explore a little, landward. But the country soon opened out into a campo, and the baked clay, uncovered with verdure, and deeply indented by the hoofs of cattle, made walking out of the question. Therefore, we were fain to turn back again, and perched upon a fence top, attempted resignation.

When the tide did turn, it made amends for all sluggishness; dashing furiously in, with a seven mile velocity, instantly flooding the shoals, and filling the channel. Quickly we were in the boat, and hurrying towards Jungcal, unaided by the paddle, save in keeping the course. The birds which had been feeding had gathered themselves hastily up, and now sat perched upon the overhanging trees, gazing down, as if they did not half comprehend the mystery of such a sudden wateriness, although daily, for their lives long, they had thus been shortened of their morning's meal. A pair of King Vultures, Urubutingas, were sailing overhead, conspicuous for their white shoulders and glossy plumage. Two miles, quickly sped, brought us to Jungcal, a small settlement of some half dozen houses, residences of the overseers and cattle drivers. We were greeted as old friends, and being just in time for breakfast, sat down—be not startled, companions of our heretofore wanderings, who have heard us discourse upon the virtues of aboriginal diet, and partaken with us of monkey and sloth, parrots, cow-fishes, and land turtles—sat down to a teak—not of exquisitely flavored victim of the Fulton Market, nor of the delicious colt-flesh of the Patagonian gourmand; but to one more exquisite, more delicious. Ah! ye young alligators, now comprehended we why chary Nature had encased ye in triple mail.

One of our objects in visiting Jungcal, was to see a rookery of Ibises and Spoonbills in the neighborhood; but as the day had so far advanced, we determined to postpone an excursion thither until the morning. Meanwhile, we amused ourselves in exploring the vicinity, and in looking over the beautiful collection of bird-skins, belonging to Mr. Hauxwell, an English collector, whom we were agreeably surprised to meet here. It was interesting to find so many of the water-birds of the United States, common here also, and to recognize in the herons, the rails, the gallinules, the ibises, the shore-birds, *et multi alii*, so many old acquaintances, in whose society we had, long ago, whiled away many a delighted hour.

Upon one side of the houses, the bamboos formed a dense hedge, but elsewhere, in every direction, stretched a vast campo, unmarked by tree or bush, save where the fringed stream but partially redeemed the general character. A few horses were feeding about, the last remnant of vast herds that once roamed the island, but which have disappeared, of late years, by a contagions pestilence; and which, judging from the specimens we saw, were any thing but the fiery coursers, described as herding on the, perhaps, more congenial plains, to the North and South.

Upon the margin of a small pond, close by, a number of Scarlet Ibises were feeding, so tame, from all absence of molestation, as to allow of near approach. Terra-terras were screaming about, and, at a distance, stalked a pair of huge white birds, known in the island as Tuyuyus, Mycteria Americana. We were exceedingly desirous to obtain one of these birds, but they were wary, and kept far beyond even rifle-shot. They are not uncommon upon the campos, and are occasionally seen domesticated in the city. A young one, which we had previously seen in the garden of the Palace, stood between four and five feet from the ground. When full grown, the Tuyuyu is upwards of six feet in height. Its neck is bare of feathers, and for two thirds of its length from above, black: the remainder is of a dark red. Its bill is about fifteen inches long, and by its habit of striking the mandibles together, a loud, clattering noise is produced. About every house were pens in which were scores of young ibises and spoonbills, which had been brought from the rookery, for the purpose of selling in Pará. They readily became tame, and well repaid the care of the negroes. Brought up for the same purpose, were parrots, paroquets, blackbirds, larks, and egrets; besides a mischievous coati, who was every where but where he should have been. Towards night, vast flocks of various water-birds came flying inland, attracting attention by their gaudy coloring and noisy flight.

Chapter XXI

DESCRIPTION OF MARAJO—CATTLE—TIGERS—ALLIGATORS—SNAKES—ANTAS—WILD DUCKS—SCARLET IBISES—ROSEATE SPOONBILLS—WOOD IBISES—OTHER BIRDS—ISLAND OF MIXIANA—INDIAN BURIAL PLACES—CAVIANA—MACAPÁ—BORE OR POROROCA—LEAVE JUNGCAL FOR THE ROOKERY—A SAIL AMONG THE TREES—ALLIGATORS—THE ROOKERY—RETURN—AN ALLIGATOR'S NEST—ADIEU TO JUNGCAL—VIOLENCE OF THE TIDE—LOADING CATTLE—VOYAGE TO PARÁ

The length of the Island of Marajo is about one hundred and twenty miles; its breadth averages from sixty to eighty. Much of it is well wooded, but far the larger part is campo, covered during the wet season with coarse, tall grass. At that time, the whole island is little more than a labyrinth of lakes. In summer, the superabundant waters are drained by numerous igaripés, and, rain rarely falling, this watery surface is exchanged for a garden of beauty, in some parts, and into a desert, upon the campos. The population of the island is large, consisting mostly of Indians and half-breeds. Some of the towns, however, are of considerable size, but most of the inhabitants are scattered along the coast and upon the igaripés. Four hundred thousand cattle roam over the campos, belonging to various proprietors, the different herds being distinguishable by peculiar marks, or brands. The estate of which Jungcal forms part, numbers thirty thousand cattle, and a great number of Indians and blacks are employed in their care, keeping them together, driving them up at proper seasons to be marked, and collecting such as are wanted for exportation to the city. These men become extremely attached to this wild life, and are a fearless, hardy race, admirable horsemen, and expert with the lasso. When horses abounded, it was customary to drive the marketable cattle towards the Pará side of the island, whence transmission to the city was easy; but, at present, they are shipped from Jungcal, or other places still more remote, thus causing great waste of time, and ruining the quality of the beef. The cattle are of good size, but not equal to those of the South. Great numbers of young cattle, and old ones unable to keep up with the herd, are destroyed by the "*tigres*," which name is applied without much precision to different species. The black tiger is seen occasionally; the Felis onça is most common of all. Neither of these is known to attack man; and in their pursuit, the islanders exhibit great fearlessness and address, never hesitating to attack them when driven to a tree, armed with a tresádo fastened to a pole. At other times, they overtake them upon the campos, running them down with horses, and lassoing them. Once thus caught, the tiger has no escape. He is quickly strangled, his legs are tied, and, thrown over the horse's back like a sack of meal, he arrives at the hut of his captor. Here a dash of water revives him, but his efforts to escape are futile. An Onça taken in this manner, was brought to Pará for Mr. Campbell. He was strangled both on being taken on and off the canoe, and after being revived, was marched upon his fore legs through the streets, two men holding each a hind leg, and others guiding him by the collar upon his neck. This animal was afterwards brought to New York by Capt. Appleton. Frequently, young tigers are exposed for sale in the market, and one of these was our fellow passenger in the Undine, upon our return. We read in works of Natural History, most alarming accounts of the fierceness of the Brazilian felines, but as a Spanish gentleman remarked to us, of the Jaguar, "those were ancient Jaguars, they are not so bad now-a-days."

The cattle have another enemy in the alligators, who seem to have concentrated in Marajo from the whole region of the Amazon, swarming in the lagoons and igaripés. There are two species of these animals, one having a sharp mouth, the other a round one. The former grow to the length of about seven feet only, and are called Jacaré-tingas, or King Jacarés. This is the kind eaten. The other species is much larger, often being seen twenty feet in length, and we were assured by Mr. Campbell, that skeletons of individuals upwards or twenty-five feet in length are sometimes encountered.

In the inner lakes, towards the close of the rainy season, myriads of ducks breed in the rushes, and here the alligators swarm to the banquet of young birds. Should an adventurous sportsman succeed in arriving at one of these places, he has but a poor chance of bagging many from the flocks around him, for the alligators are upon the alert, and the instant a wounded bird strikes the water, they rush en masse for the poor victim, clambering over one another, and crashing their huge jaws upon each others' heads in their hasty seizure. Late in the wet season, they lay their eggs, and soon after, instead of becoming torpid, as would be the case in a colder climate, bury themselves in the mud, which, hardening about them, effectually restrains their locomotion, until the next rains allow their dislodgment. The inhabitants universally believe, that the alligator is paralyzed with fear at the sight of a tiger, and will suffer that animal to eat off its tail, without making resistance. The story is complimentary to the tiger, at all events, for the tail of the alligator is the only part in esteem by epicures.

Snakes spend their summers in the same confinement as alligators, and upon their issuing forth, are said to be very numerous, and often of great size. It was from Marajo, that the anaconda, now or lately exhibited at the American Museum, was brought, and this fellow, as well as the "Twin Caffres," we frequently saw at Pará before their transportation to New York. The largest snake known, of late years, at Pará, was twenty-two feet in length. He was captured upon Fernando's Island, near the city, by the negroes, with a lasso, as he laid upon the shore, basking in the sun. He had long infested the estate, carrying off, one time with another, about forty pigs. Even after being captured and dragged a long way to the house, he coiled his tail around a too curious pig, that we may suppose, was gloating over his fallen enemy, and would have made a forty-one of him, had not the exertions of the blacks forced him to let go his hold.

We never heard an instance of snakes attacking man, and the negroes do not fear an encounter with the largest. Snake hunts, doubtless, have exciting interest, as well as others less ignoble. As elsewhere remarked, these reptiles are very frequently kept about houses in the city, and may be often purchased in the market, nicely coiled in earthen jars. Southey records an old story to this effect: "that when the anaconda has swallowed an anta, or any of the larger animals, it is unable to digest it, and lies down in the sun till the carcass putrefies, and the urubus, or vultures, come and devour both it and the snake, picking the flesh of the snake to the back-bone, till only back-bone, head and tail be left; then the flesh grows again over this living skeleton, and the snake becomes as active as before." The march of knowledge in this department is certainly onward; *now*, gentlemen in Pará believe no more, than that the whole belly and stomach fall out, trap-door like, soon to heal again, and ready for a repetition. In either case, the poor snake is much to be pitied.

The Antas, or Tapirs, are animals not often found upon the main-land, but occasionally observed on Marajo, along the igaripés. They are, by many, considered as amphibious, but they live upon the land, merely resorting to the water for bathing. In size, they resemble a calf of a few months, and when old, are of a brown color. They are remarkable for a proboscis-like nose. When tamed, they are extremely docile, and are allowed to roam freely, being taught to return home regularly. One which we saw in this estate was small, and marked with longitudinal spots of a light color.

The large Ant Eater is also a dweller on Marajo.

The Ducks breeding upon this island are of two kinds, the common Musk Duck, and the Maracas (Anas autumnalis). The latter are most numerous. By the month of September, the young are well grown, and the old birds are debilitated from loss of their wing quills. Then, particularly upon Igaripé Grande, on the Pará side, people collect the ducks in great flocks, driving them to a convenient place, and, catching them, salt them down by the canoe load.

Of the water birds frequenting Marajo, the Scarlet Ibis, and the Roseate Spoonbill, excel all in gorgeousness and delicate coloring. The Ibises are of the brightest scarlet, excepting the black tips of the wings, and their appearance, when, in serried ranks, the length of a mile, they first come to their breeding place, is described, as one might well imagine it, as wonderfully magnificent They appear, in this manner, in the month of June, and, at once, set about the forming of their nests. At this time, they are in perfect plumage, but soon commencing to molt, they lose somewhat of their beauty. The young birds are ready to depart in December, and then, the whole family disappear from the vicinity, excepting a few individuals here and there. In Maranham, the breeding season is in February, and, in that month, Capt. Appleton found them there in vast numbers. Sometimes, but rarely, they are observed in the Gulf districts of the United States, but they have never been known to breed there. The nests are made of small sticks, loosely formed. From two to three eggs are laid, greenish in color, and spotted with light brown.

The Roseate Spoonbills do not migrate as do the Ibises, being quite common upon the whole coast, and sometimes being seen far up the Amazon in summer. The delicate roseate of their general coloring, with the rich, lustrous carmine of their shoulders, and breast tufts, as well as the singular formation of their bills, render them objects of great interest as well as beauty. They are seen fishing for shrimps and other small matters along the edges of the water, or in the mud left exposed by the ebbing tide, and as they eat, grind the food in their mandibles moved laterally. As well as the Ibis, they are exceedingly shy at every season, except when breeding. They breed in the same places with the Scarlet Ibises and the Wood Ibises, and the nests of the three resemble each other in every respect, but in size. The eggs of the Spoonbill are from three to four, large, white, and much spotted with brown. The birds are called by the Brazilians, Colheréiros, meaning spoonbill. The name of the Ibis is Guerra, signifying warrior.

Another of the northern birds here breeding, is the Wood Ibis, Tantalus loculator, much larger than either of the above. Its general plumage is white, the tips of the wings, and the tail, being a purplish black. By the natives, it is called the Jabirú, which name in Ornithologies is more generally applied to the Tuyuyu. It lays two or three eggs, of a dirty white color.

Besides these, the Glossy Ibis, Ibis falcinellus; the Great Blue Heron, A. herodias; Night Heron, A. nycticorax; Great American Egret, A. alba; Snowy Heron, A, candidissima; Least Bittern, A. exilis; Purple Gallinule, Black-necked Stilt, and perhaps others common in the United States breed upon Marajo; as well as a variety of the same family peculiar to the South.

We found here, also, one of the rarer land-birds of Audubon, the Fork-tailed Fly-catcher, Muscicapa forficatus, and were fortunate enough to discover its nest. This was near the water, in a low tree, and was composed of grass and the down of some plant. The eggs were two in number, white, and spotted with brown, at the larger end more particularly, resembling, except in size, those of our King-bird.

Generally, the land-birds upon Marajo are of different varieties from those found about Pará, and upon the Main. The Chatterers are not seen there; the Toco Toucan takes the place of the Red-billed; the Cayenne Manikin, whose head and shoulders are bright red, is as common as the White-capped elsewhere; Black-backed Yellow Orioles, Icterus jububu, are extremely abundant; as are also the Mango Humming Bird, T. mango; the Ruby and Topaz, T. moschitus; Swallow-tailed, T. forficatus; Black-breasted, T. gramineus; and many other varieties of this family.

Opposite Jungcal, and in view from the shore, is the Island of Mixiana, twenty-five miles in length, and resembling Marajo in its characteristics. This is entirely the property of Srs. Campbell and Pombo, the proprietors of the Jungcal estate, and here they have many thousand cattle.

Upon Mixiana are Indian burial places, and from these are disinterred urns of great size, containing bones and various trinkets. Unfortunately, our time would not allow us to visit that island, or we should have been at the pains of exploring these interesting remains. We saw, however, one of the jars at Jungcal. Similar burying places are found in various parts of Brazil and Paraguay, and the ancient method of interment in most of the tribes was the same.

Beyond Mixiana, is the much larger island of Caviana, and many other islands, of considerable size, are strewn over the mouth of the river.

Upon the opposite shore is the town of Macapá, said to contain the finest fort in Brazil. The situation is considered unhealthy, and foreigners rarely visit there. Sailing from Pará to Macapá, one passes more than forty islands. Between Macapá and Marajo is seen in its perfection the singular phenomenon, known as the Bore, or Pororoca, when the flood tide, at the instant of its turning, rolls back the waters of the river in an almost perpendicular wall. Condamine, many years ago, described the sea as "coming in, in a promontory from twelve to fifteen feet high, with prodigious rapidity, and sweeping away every thing in its course." No one knows of such terrible phenomena now-a-days. We inquired of several persons accustomed to piloting in the main channel, and of others long resident in the city and familiar with the wonders of the province, but none of them had known the water to rise above the height of five feet, even at the spring tides. A canoe of any size is in no danger, her bow being turned to the flood.

Early in the morning, we accompanied Mr. Hauxwell to a tree, upon which a pair of Tuyuyus were building their nest. A nimble Indian climbed the tree, but the nest was unfinished. It was thirty feet from the ground, composed of large sticks; and looked from below, big enough for the man to have curled himself in.

We left Jungcal for the rookery, about nine o'clock, with the flood tide, in a montaria, with a couple of guides. They were men of the estate, and looked upon the adventure as most lucky for them. Making pleasure subservient to business, they carried their harpoons for fish or alligators, and baskets for young birds.

Immediately after leaving the landing, we startled a Cigana from her nest, in the low bushes by the water. The stream grew more and more narrow, winding in every direction. Tops of tall trees met over our heads, countless flowers filled the air with perfume, and the light and shade played beautifully among the green masses of foliage.

Upon the trees were perched birds of every variety, who flew before our advance, at short distance, in constantly increasing numbers, or curving, passed directly over us; in either case, affording marks too tempting to be neglected. Upon some topmost limb, the Great Blue Heron, elsewhere shyest of the shy, sat curiously gazing at our approach. Near him, but lower down, Herons, white as driven snow; some, tall and majestic as river naiads, others, small and the picture of grace, were quietly dozing after their morning's meal. Multitudes of Night Herons, or Tacarés, with a loud quack, flew startled by; and, now and then, but rarely, a Boatbill, with his long plumed crest, would scud before us. The Snake-bird peered out his long neck, to discover the cause of the general commotion; the Cormorant dove from the dry stick, where he had slept away the last hour, into the water below; swimming with head scarcely visible above the surface, and a ready eye to a treacherous shot. Ducks rose hurriedly, and whistled away; Curassows flew timidly to the deeper wood; and fearless Hawks, of many varieties, looked boldly on the danger.

With a noise like a failing log, an alligator would splash into the water from the bank, where she had been sunning herself, or looking after her nest; and often, at once, half a dozen huge, unsightly heads were lifted above the surface, offering a fair, but not always practicable mark for a half-ounce ball. Occasionally, a whole family of little alligators, varying in length from six to eighteen inches, would start out of the leaves instinctively, some, plumping themselves in, as the examples of their respected mammas had taught them; others, in their youthful innocence, standing gazing at us, from the top of the bank; buy with more than youthful cunning, ready also to plump in at the least motion towards raising a gun. At frequent intervals, the beaten track from the water, disclosed the path of some of these monsters; and a pile of leaves, just seen through the trees, showed clearly the object of their terrestrial excursions.

As we neared the rookery, after a two hours' pull, the birds were more and more abundant; and the alligators more and more bold, scarcely minding our approach, and only learning caution, by repeated applications of leaden balls. The frequent proximity of the King Jacarés, offered many opportunities, to the harpooner in the bow; but we learned, by his ill success, that these autocrats cared very little for punches in the ribs.

Turning suddenly, we left the bordering forest for a canebrake, and instantly broke full upon the rookery. In this part, the Scarlet Ibises, particularly, had nested; and the bended tops of the canes were covered by half-grown birds in their black plumage, interspersed with many in all the brilliance of age. They seemed little troubled at our approach, merely flying a few steps forward, or crossing the stream. Continuing on, the flocks increased in size; the red birds became more frequent, the canes bent beneath their weight like reeds. Wood Ibises and Spoonbills began to be numerous. The nests of all these, filled every place where a nest could be placed; and the young Ibises, covered with down, and standing like so many Storks, their heavy bills resting upon their breasts and uttering no cry, were in strong contrast to the well-feathered Spoonbills, beautiful in their slightly roseate dress, and noisily loquacious. Passing still onward, we emerged from the canes into trees; and here the White Herons had made their homes, clouding the leaves with white. Interspersed with these, were all the varieties mentioned before, having finished their nesting, and being actively engaged in rearing their young. We had sailed above a mile, and at last, seeming to have approached the terminus, we turned and went below a short distance to a convenient landing, where we could pursue our objects at leisure. The boatmen, at once, made their dispositions for basketing the young birds; and soon, by shaking them down from the nests, and following them up, had collected as many as they desired. We wandered a long distance back, but the nests seemed, if any thing, more plentiful, and the swarms of young more dense. At the sound of the gun, the birds in the immediate vicinity, rose in a tumultuous flock; and the old ones circled round and round, as though puzzled to understand the danger they instinctively feared. In this way, they offered beautiful marks to oar skill; and the shore, near the canoe, was soon strewed with fine specimens. Evidently, this place had been for many years, the haunt of these birds. Not a blade of grass could be seen; the ground was smooth and hard and covered with excrement.

Occasionally, and not very rarely, a young heedless would topple into the water, from which the noses of alligators constantly protruded. Buzzards, also, upon the bank; sunned themselves and seemed at home; and not unfrequently, a hungry Hawk would swoop down, and away with his prey almost unheeded.

We were amused by the manner of feeding the young Scarlet Ibises. In the throat of the old female bird, directly at the base of the lower mandible, is an enlargement of the skin, forming a pouch, which is capable of containing about the bulk of a small hen's egg. She would return from fishing on the shallows, with this pouch distended by tiny fish, and allowed her young to pick them out with their bills.

It was late when the tide turned and we hastened away, with the canoe loaded to overflowing. The birds seemed now collecting for the night, Squads of bright-colored ones were returning from the shore, and old and young were settling on the canes, over the water, like swallows in August. An alligator gave us an opportunity for a last shot, and the air was black with the clouds of birds that arose, shrieking and crying. I never conceived of a cloud of birds before.

On our way down, we discovered the nest of a Socco, the Tiger Bittern, close by the water. The old bird observed our motions for an ascent with indifference, when, up through the feathers of her wing, peered the long neck of a little fellow, intimating that we might as well be off; for it was of eggs we were greedy.

Soon after, we arrived at the spot, which we had marked in the mrorning, where an alligator had made her nest, and *sans ceremonie*, proceeded to rifle it of its riches. The nest was a pile of leaves and rubbish, nearly three feet in height, and about four in diameter, resembling a cock of hay. We could not imagine how or where the animal had collected such a heap, but so it was; and, deep down, very near the surface of the ground, from an even bed, came forth egg after egg, until forty-five had tolerably filled our basket. We kept a good look-out that the old one did not surprise us in our burglary, having read divers authentic tales of the watchful assiduity of the mother. But nothing appeared to alarm us, and we concluded, that, like others of the lizard family, alligators are merely anxious to make their nests, and trust to the fermented heat, and to Providence, for hatching and providing for their brood of monsters. These eggs are four inches in length, and oblong; being covered with a crust rather than a shell. They are eaten, and our friends at the house would have persuaded us to test the virtues of an alligator omelet, but we respectfully declined, deeming our reputations sufficiently secured by a breakfast on the beast itself.

Ave Maria had sounded when we reached Jungcal, and the satisfaction we felt at the close of this, the greatest day's sporting we had ever known, amply compensated for all our fatigue. The boat in which we came being obliged to return immediately, we were under the necessity of leaving this delightful spot, where we could have been content to while away a month. But one such day as we had passed, repaid us for the inconveniences of a week upon the water.

We bade adieu to our good friends in the morning, taking the last of the ebb to arrive at the vessel. But when quite near, the tide turned, the flood rushed in, and we were very likely to revisit Jungcal. However, by running in shore, and claiming assistance of the overhanging canes, after a weary pull, we reached our goal, almost inclined to credit M. Condamine.

The crew were loading with the cattle, which had been driven down the day before, and were now confined in the ken. This was enclosed on every side; but that toward the water. A dozen men stood inside and out, some holding the lasso, others ready to pull, the instant the animal was caught, and others, still, were armed with sharp goads, with which to force him onward. Some of the cattle showed good Castilian spirit, and their rage was several times with difficulty eluded by a leap to the friendly fence. Once in the water, their struggles were over. A rope was fastened about their horns, and thus they were hoisted up until they were above the hole in the deck made to receive them. Below, they were secured to side beams, and were scarcely allowed room to move.

Putting out of the igaripé, for two days we were beating to windward, anchoring half the time, and being tossed about in a way to make us curse all cattle boats. The poor victims in the hold fared worse than we, deprived of food and drink, pitched back and forth with every motion, and bruised all over by repeated falls upon the rough floor. We lost all gusto for Pará beef. From Cape Magoary we had a fine run, reaching Pará upon the third night.

Chapter XXII

WANT OF EMIGRANTS AND LABORERS—INDUCEMENTS TO SETTLERS, AND DISADVANTAGES—CITIZENSHIP—IMPORT AND EXPORT DUTIES AND TAXES—WANT OF CIRCULATING MEDIUM—EMBARRASSMENTS OF GOVERNMENT—CAPABILITIES OF THE PROVINCE—EFFECT OF CLIMATE ON THE WHITES—THE BLACKS—INDUCEMENTS TO THE FORMATION OF A STEAMBOAT COMPANY—SEASONS—TEMPERATURE—HEALTH—SUPERIOR ADVANTAGES TO INVALIDS—FAREWELL TO PARÁ—VOYAGE HOME

The want of emigrants from other countries, and of an efficient laboring class among its population, are the great obstacles to the permanent welfare of Northern Brazil. It never was the policy of Portugal to encourage emigration excepting from her own territory, and although, by the indomitable enterprise of her song, she secured to herself the finest Empire in the world, yet for want of other assistance, this Empire is impoverished, and the millions of square miles that should now be teeming with wealth, are entirely unproductive. With the nobler qualities of the old Portuguese, to which popular history has never done justice, was mingled s narrowness of mind, that was natural enough in the subjects of an old and priest-ridden monarchy. The Brazilians have not entirely thrown off this prejudice of their ancestors, and still entertain somewhat of the old jealousy of foreigners, but very naturally, in a newly liberated government, they dislike the Portuguese above all others. Much of the wealth of the country is in the hands of the Portuguese, who, coming over when young, with habits of shrewdness, and economy, almost always accumulate fortunes. The Brazilians are no match for them in these qualities, and therefore hate them most cordially. For the same reason, this feeling is continually excited, although in a lesser degree, against other foreigners, but more in some parts of the Empire than others, and probably, as little in Pará as any where.

The Brazilian government offers great inducements to emigrants, and yet these are more than neutralized by disabilities and present disadvantages. Land is free of cost, and upon any vacant section, a man may settle, with the proprietorship of, at least, a square league, and as much more as he really requires. Moreover, any new improvement in tools, or machinery, may be introduced free of duties.

The ground is easily cleared, as the roots of the trees do not extend far beneath the surface, and the efforts of man are further aided by causes attendant upon the clime. The soil is of the greatest fertility, and sugar cane, rice, coffee, anatto, cotton, cacao, and a hundred other products, richly repay the labor bestowed upon their cultivation; while from the forests are obtained gums and drugs, all yielding a revenue. Almost every thing grows to hand that man requires; living is cheap, and the climate delightful.

On the other hand, the counteracting obstacles are very great. Although the government professes every desire for the accession of foreigners, it denies them the rights of citizenship, excepting under peculiar circumstances, which, of course, obliges them to labor under legal disabilities.

Again, import duties are extravagantly high, and articles of furniture, tools, or machinery, which cannot be manufactured in the country without great expense, if at all, are taxed so highly as to be really prohibited; although, as before stated, new inventions and improvements, are introduced from abroad without charge.

But a greater drawback, by far, is the export duty, the most stupid, indefensible measure that could be conceived; a withering curse to all enterprise, and a more effectual hindrance to the prosperity of Brazil, than a weak government, dishonest officials, a debased currency, and all other influences together. Brazilian statesmen imagine that the export tax comes directly from the pocket of the foreign purchaser, whereas, it recoils upon the producer, and its effect is to make the price paid for labor so low, as to prohibit cultivation. There is scarcely a product raised in the two countries, in which Brazil could not undersell the United States in every market of the world, were it not for this tax. Its cotton and rice, even during the past year, have been shipped from Pará to New York. Its tobacco is preferable to the beat Virginian, and can be raised in inexhaustible quantities.

The imposition upon the producer is also increased by the tithe required for the church, and, between the two, the lower classes are under a burden, which occasionally becomes insupportable, and which is the undoubted cause of the general and increasing disaffection toward the government, and of the revolutions which have heretofore broken out, and which are always feared. Rubber shoes, which are principally made by the low whites and Indians, pay three taxes to the treasury before they leave the country, until the first price is nearly doubled. Not a basket of oranges, or of assai, comes to market, untaxed.

Not only do products exported to foreign countries pay duties, but even from one Brazilian port to another, and from one inland town to another. A few bags of coffee, which were sent by us from the Barra of the Rio Negro to Santarem, paid duties at the latter place. Chili hats, coming from Peru, pay duties at the frontier, again at Pará, and again at Rio Janeiro. No country in the world could bear up under such intolerable exactions, and Brazilian statesmen may thank their own folly if the Empire be dismembered.

Another obstacle, severely felt, is the want of a circulating medium. The Brazilian currency consists almost entirety of copper, and paper issued by the government. The smallest value is one ree, corresponding to one half mill in our currency, and the smallest coin is of ten rees: the largest of eighty; or four vintens. One thousand rees make a milree, the smallest paper note, about equal in value to a half dollar. There are various issues, from one milree to one thousand. Excepting in the city, and upon the remote frontiers, gold and silver will not circulate. The amount of bills, in the province of Pará, is never adequate to the wants of the people, and their tendency is always to the city. Furthermore, by the operations of government, even the little currency that is floating, is constantly fluctuating in value. Upon one pretext or another, they call in notes of a certain denomination, at short notice, and under a heavy discount. Such was the case with the two milree notes, when we were upon the river. Not long since, it was discovered that the Treasurer at Rio Janeiro, had sent to the provinces a vast amount of money for the payment of the troops, which was certainly struck off the original plate, but differed from the true emission by the absence of a letter or word. It was a fraud of the Treasurer, unless, as many believed, sanctioned by the government. These bills were scattered to the remotest corners of the Empire, when suddenly appeared an order, recalling the whole, within a certain limited time. If this were a speculation of the government, it was, probably, a profitable one, though the country may not have received the benefit of it. But a few years since, one milree was nearly or quite equivalent in value to one dollar in silver.

The truth is, that the Brazilian government is a weak government. It is too republican to be a monarchy, and too monarchical to be republic. If it were decidedly one or the other, there would be greater strength and greater freedom; but now, it has neither the bulwark of an aristocracy, nor the affection of the people. It is forced to depend entirely upon a regular army for its existence, and is kept in a state of constant alarm by disturbances in its provinces, or invasions of its frontiers; it is bowed beneath a heavy foreign debt, and obliged to use all kinds of expedients, not to make advance, but to retain its position.

Were Pará a fiee and independent State, its vast wilds would, in a few years, be peopled by millions, and its products would, in a few years, be peopled by millions, and its products would flood the world. It contains an area of 950,000 square miles, nearly half the area of the United States and all its territories. Its soil is every where of exhaustless fertility, and but an exceedingly small portion of it is unfitted for cultivation. The nobles rivers of the world open communication with its remotest parts, and lie spread like net-work over its surface. It is estimated that the Amazon and its tributaries present an aggregate navigable length of from 40,000 to 50,000 miles. The whole territory is as much superior, in every respect, to the valley of the Mississippi, as the valley of the Mississippi is to that of the Hudson.

But besides the hindrances to prosperity on the part of the government, the settler has other disadvantages to struggle against, one of which being the deficiency of means of transportation throughout the interior, may be but temporary; the other is the effect of the climate. It is not to be denied, that although the climate is singularly healthy, its constant heat is enervating, and that natives of colder regions, after few years' residence, have not that bodily strength requisite to daily and protracted toil. It is only in the early morning, and late in the afternoon, that white men can labor in the open air; but where a white would inevitably receive a sun-stroke, a negro labors with uncovered head, without injury or exhaustion. The one has capacity to direct, and the other the ability to perform, and it is difficult to conceive how the resources of Brazil can ever be successfully developed, without a co-operation of the two races. The blacks need not be slaves, they would answer every purpose, in being apprentices after the British West India system.

Brazilian slavery, as it is, is little more than slavery in name. Prejudice against color is scarcely known, and no white thinks less of his wife because her ancestors came from over the water. Half the officers of the government and of the army, are of mingled blood; and padres, and lawyers, and doctors of the intensest hue, are none the less esteemed. The educated blacks are just as talented, and just as gentlemanly as the whites, and in repeated instances, we received favors from them, which we were happy to acknowledge.

Efforts have been made for the establishment of steamboats upon the Amazon, but from causes unforeseen, and not inherent in the enterprise, they have failed. A fen yean since, the government granted a monopoly of the river, for a term of years, to a citizen of Pará. A company was formed, and a small steamboat brought out, but from lack of confidence in the individual referred to, the enterprise progressed no further. It is said, the government are ready to renew their offers, and there can be no question but that an efficient company would meet success. Such a company should have sufficient capital to enable it to purchase its own freight in the interior, at least, in the beginning of the enterprise. For, at first, the novelty of the thing, and the general dislike to innovation, would prevent the co-operation of the people at large. Time and success would soon wear away their prejudices. The present method of transportation is so tedious and expensive, that a steamboat would destroy all opposition from the river craft, and by appointing proper agents in the several towns, and making the upper depot at the Barra of the Rio Negro, constant and profitable freights would always be secured.

A boat built of the wood of the wood of the country, would be preferable, on account of its not being affected by boring worms in the water, or by insects; but perhaps the former might be avoided by copper.

The navigation of the river is perfectly clear, excepting in the Bays of Marajo and Limoeiro, and surveys in these, would no doubt discover convenient channels. There are neither snags, nor sawyers; the only thing of the kind being floating cedars, easily guarded against.

If a company were formed, much of the stock would be taken in Pará, and the enterprise would receive every encouragement from the citizens. Sooner or later, the Amazon must be the channel of a vast commerce, and Pará must be, from the advantages of its situation, one of the largest cities of the world.

It remains further to speak of the climate of Pará, and of the extraordinary advantages which it present to invalids and travelers.

The seasons are, properly speaking, but two, the rainy and the dry. The former commences about the 1st of January, and continues until July. During the first part of this time, rain pours unremittingly; then, for a season, the greater part of the afternoon and night, and, at last, perhaps only in a daily shower. At this time, also, the trade-winds blow with less regularity than in summer.

Throughout the dry season, more or less rain falls weekly, but strong trades blow, heavy dews distil, and the climate is perfectly delightful. This season commences, in the interior, one or two months earlier than at Pará, and, during its continuance, rain falls more rarely. At this time, a passage up the river is speedy, and a descent exceedingly tedious. Senhor Henriquez told us, that he was once sixty days in coming from the Rio Negro to Pará, in a small boat, on account of the winds. Thunder and lightning rarely accompany the rains, and any thing approaching a tornado is almost unknown.

It seems singular, that directly under the equator, where, through a clear atmosphere, the sun strikes vertically upon the earth, the heat should be less oppressive than in the latitude of New York. This is owing to several causes. The days are but twelve hours long, and the earth does not become so intensely heated as where they are sixteen. The vast surface of water constantly cools the air by its evaporation, and removes the irksome dryness, that in temperate regions, renders a less degree of heat insupportable. And finally, the constant winds blowing from the sea, refresh and invigorate the system.

According to observations made by Mr. Norris, during the month of June, July and August, at the hours of 6 A. M., 3 P. M., and 8 P. M., the mean temperature for June was 79° 98′ Far.; the highest 86°, lowest 77°: for July, the mean was 80° 54′; highest 86°, lowest 77°: for August, the mean was 80° 92′; highest 86°, lowest 77°. The mean for the three months was 80° 48′, and the variation but 9°. I do not believe that another spot upon the face of the earth can show a like result. This heat we never felt to be oppressive, except when dining in state, in black cloth coats. Moreover we were never incommoded by heat at night, and invariably slept under a blanket. The reason for this, and also for wearing flannel next the skin, at all times, is, that in a very few weeks, a person becomes so acclimated as to be sensitive to a very slight degree of variation in the temperature.

This equality of temperature renders the climate Pará peculiarly favorable to health. There is no kind of epidemic disease; people live to a good old age, and probably the average of life is as high as in the city of New York.

Such a climate is invaluable to invalids, particularly those suffering from pulmonary complaints. Two hundred years ago, Sir William Temple wrote alter this manner upon the Brazilian climate generally: "I know not whether there may be anything in the climate of Brazil more propitious to health, than in other countries; for besides what was observed among the natives upon the first European discoveries, I remember Don Francisco de Mello, a Portugal ambassador in England, told me, it was frequent in his country for men spent with age or other decays, so as they could not hope for above a year or two of life, to ship themselves away in a Brazil fleet, and upon their arrival there, to go on to a great length, sometimes of twenty or thirty years, or more, by the force of that vigor they received with that remove. Whether such an effect might grow from the air, or the fruits of that climate, or by approaching nearer the sun, which is the fountain of life and heat, when their natural heat was so far decayed; or whether the piecing out of an old man's life were worth the pains, I cannot say". This is more true of the climate of Pará, than of any other part of Brazil.

Multitudes of persons from the Northern States, now visit the south, in search of health, or spend their winters in the West India Islands, at great expense, and little gain, who in Pará, could reside for comparatively nothing, with a certainty of recovery. The passage out is low, from fifty to seventy-five dollars, and living in the city is cheap. At present, there are no houses for public accommodation, but until the influx of strangers imperatively required one, the citizens and the foreign residents would receive the comers with open arms. And Brazilian hospitality is not hospitality only in name; it is the outflowing of a noble and generous warm-heartedness that would redeem a thousand failings. But if individuals prefer, houses are always to be obtained and servants always to be hired, and they may live as they please.

The novelty and beauty of the country, as well as the luxury of the climate, afford sufficient inducements to the invalid for seeking both health and pleasure, in Pará, while its trees and flowers, birds, shells and insects, offer exhaustless resources for diverting the mind, and promoting the bodily exercise necessary to a recovery of health.

Good medical care is always present; the physicians of the city being graduates from European universities. Moreover, the medicines peculiar to the country are of great number and efficacy, and there is scarcely a form of disease for which Nature has not a remedy at hand. An instance in point came directly under our observation, the gentleman who was the patient being for several weeks with us, at the house of Mr. Norris. He had gone out from the United States with his system so filled with mercury, that his mouth was ulcerated, his teeth dropping out, and his joints so affected that every motion produced agony. He was recommended, at Pará, to try a remedy called by the Indians Mu-lu-ré, which is the juice of a creeping plant found plentifully throughout the country. In three weeks, our friend was perfectly cured, and is now in the United States, a well man. We heard of similar astonishing cures from other individuals who had been the subjects, and every one in Pará is acquainted with the virtues of the medicine. Why it has not been known abroad, it is difficult to say.

There is a wide field for medical inquiry yet left in the Brazilian forests, and one that demands to be explored.

It may be that some naturalist or sportsman may be incited by the recent accounts of adventures on the Amazon, to undertake an expedition thither for research or pastime, and as we ourselves were unable to gain proper information with regard to the articles necessary to an outfit, a few words upon that subject will, perhaps, not be useless. In the way of clothes, half a dozen suits of light material, some of which are calculated for forest wear, are necessary, and may be obtained ready made, and at low prices, at any of out Southern clothing stores; as well as check and flannel shirts. A black dress suit is required by Pará etiquette. A naturalist's implements must also be taken out, as well as powder, fine shot, arsenic, flower presses, and paper and wooden boxes for insects and other objects. Many of these things cannot be obtained at all, or only at extravagant prices and of poor quality, at Pará.

As for medicines, we took out a well filled chest, and excepting for one or two doses of calomel, never opened it on our own account. Hartshorn is more valuable than aught else, being effectual against the stings of all insects.

Hammocks are always to be had, but blankets are not, and if a man intends to stretch himself upon hard boards, a rubber pillow is rather softer than a gun-case. We also took out a variety of rubber articles. The clothes' bags were useful, and the light cloaks answered in the absence of something better, but as a general thing, the articles were all humbugs. And most especially are rubber boots, which ought to have been known to the Inquisition. A far better article for a cloak is the Spanish poncho, a square cloth, with a hole in the middle, for the neck. Made of heavy cloth, and lined with baize, no rain since the deluge could wet it through, and it always answers for bed or pillow.

As to ignorance of the language, that is a matter of no consequence. The Portuguese is intimately allied to the Spanish, and is one of the most easily acquired languages in the world. A stranger readily learns the necessary phrases, when he is compelled to do so, and a few weeks' attention renders him sufficiently on adept for all practical purposes. Not only are there many foreigners in Pará who speak English, but it is very generally understood by the Brazilian and Portuguese merchants of the city.

It was a delightful morning in the latter part of October, when, in the good bark Undine, we bade adieu to Pará. We had come from winter into summer, and were now returning to winter again, and although the thoughts of home were pleasant, it was very hard to part with kind friends, and to say a farewell, that was to be perpetual, to this land of sunshine, of birds and flowers.

Our passage was long and tedious. For days, we lay becalmed beneath torrid burnings, and when winds did come, they blew in furious gales. But we had wherewithal to amuse ourselves, and upon sundry occasions, enlivened the mornings by spearing a dolphin, or by hooking a shark. The parrots and monkeys, too, exerted themselves in our behalf. Some of the parrots died, and the prized gift of Senhor Bentos deliberately dove from one of the upper yards, into the deep, deep sea. The paroquets bore the voyage bravely, housed in a flannel-covered basket, and yellow-top now chatters as merrily as in his far distant home, by the Rio Negro. The little duck that we picked up from the water, under the Christian designation of Paddy, swims proudly in an Ulster lake, and discourses to the marakong geese who keep him company, of the sudden changes of life, and the virtue of contentment. But the poor macaw, who had our faithful companion from the remotest point of our travels, and who had made a triumphant entry into New York streets, covered in a blanket, and declaiming lustily to passers-by, ventured, one cold night, to the outer yard, and perished the victim of his imprudence.

THE NARRATIVE PRESS
TRUE FIRST-PERSON HISTORICAL ACCOUNTS

THE HISTORICAL ADVENTURE AND EXPLORATION SERIES

The *Historical Adventure and Exploration Series* from The Narrative Press are all first-hand reports written by the explorers, pioneers, scientists, mountain men, prospectors, spies, lawmen, and fortune hunters themselves.

Most of these adventures are classics, about people and places now long gone. They take place all over the world – in Africa, South America, the Arctic and Antarctic, in America (in the Old West and before), on islands, and on the open seas.

Some of our authors are famous – Ernest Shackleton, Kit Carson, Henry Stanley, David Livingston, William Bligh, John Muir, Richard Burton, Elizabeth Custer, Teddy Roosevelt, Charles Darwin, Osborne Russell, John Fremont, Joshua Slocum, William Manley, Tom Horn, Philip St. George Cooke, Apsley Cherry-Garrard, Richard Henry Dana, Jack London, and Buffalo Bill, to name a few.

One thread binds all of our books: every one is historically important, and every one of them is fascinating.

Visit our website today. You can also call or write to us for a free copy of our printed catalogue.

The Narrative Press
P.O. Box 2487
Santa Barbara, California 93120 U.S.A.
(800) 315-9005
www.narrativepress.com